TSUNAMI

The Earth series traces the historical significance and cultural history of natural phenomena. Written by experts who are passionate about their subject, titles in the series bring together science, art, literature, mythology, religion and popular culture, exploring and explaining the planet we inhabit in new and exciting ways.

Series editor: Daniel Allen

In the same series
Desert Roslynn D. Haynes
Earthquake Andrew Robinson
Fire Stephen J. Pyne
Flood John Withington
Islands Stephen A. Royle
Moon Edgar Williams
Tsunami Richard Hamblyn
Volcano James Hamilton
Waterfall Brian J. Hudson

Tsunami

Richard Hamblyn

REAKTION BOOKS

Published by
Reaktion Books Ltd
33 Great Sutton Street
London EC1V 0DX, UK
www.reaktionbooks.co.uk

First published 2014

Printed and bound in China by 1010 Printing International Ltd

A catalogue record for this book is available from the British Library

ISBN 978 1 78023 347 5

CONTENTS

Preface: The Tsunami Stone

The sea has neither meaning nor pity.[1]

Dotted along Japan's northeast coast stands a series of weather-beaten obelisks, each engraved with the following warning: 'Remember the calamity of the great tsunamis. Do not build any homes below this point.' The stones, some of which have stood for nearly 600 years, are testament to Japan's long exposure to the dangers of tsunamis, and – as many coastal communities have learned over the years – the advice they offer can save lives. The inhabitants of Aneyoshi, for example, a hillside village in Iwate Prefecture, credit their local warning stone for preserving them from the catastrophic tsunami of March 2011.[2] Aneyoshi stands immediately above the point where the 2011 tsunami reached its greatest height: a mountainous 38.9 m (128 ft), a record now marked by a blue line painted on the road. Below the line the once wooded valley remains a scene of devastation.

It is no coincidence that the same location was the site of Japan's previous tsunami run-up record: 38.2 m (125 ft), dating from the 1896 Meiji Sanriku tsunami. The particular coastal topography around that part of northeast Honshu can lead to the rapid shoaling and funnelling of tsunamis and storm surges, creating unusually high waves along Iwate's rocky shores. It was the 1896 tsunami, with its death toll of more than 27,000, which first convinced the surviving inhabitants of Aneyoshi to move their village up the valley. But within a generation or two the population had recovered and the villagers began to build new houses further down the hill, closer to their fishing harbour. In March 1933, however, another deadly tsunami crashed into the

A memorial stone in the village of Aneyoshi, Iwate Prefecture, Japan: 'High dwellings are the peace and harmony of our descendants. Remember the calamity of the great tsunamis. Do not build any homes below this point.'

overleaf: The black wave breached the sea defences of Miyako City, northeast Japan, on 11 March 2011. Nearly 16,000 people died in the tsunami across Japan.

same unlucky section of coast, this time leaving only four villagers alive.

It was they who erected the warning stone, and since then Aneyoshi has remained above the waterline, safe from subsequent Pacific tsunamis – even the vast implacable monsters that surged ashore in May 1960 and March 2011. But many other coastal villages had ignored the warnings on their stones, regarding them as relics from a bygone age, and had rebuilt too close to the water. The fact that several of these overlooked warning stones were swept away by the 2011 tsunami, along with the villages they were designed to protect, suggests that modern confidence in sea walls and warning systems can be tragically misplaced, and that costly technology is not always a match for the hard-won wisdom of the past.

Maybe we should learn to listen to the stones again, and take their lessons to heart; for, as every chapter of this book will show, it is stories and memories that save lives.

1 Tsunamis in History and Memory

> This wave just got bigger and bigger. That was the first time that anybody around us, anybody, thought to be afraid. Here we were, landlubbers, and it never occurred to anyone to be afraid.[1]

One of the most enduring stories inherited from antiquity is the legend of the drowned civilization of Atlantis. It first appeared in Plato's Socratic dialogues *Timaeus* and *Critias* (composed *c.* 360 BC), and for centuries this story of a wealthy island empire destroyed by the gods was assumed to have been a fable, a cautionary tale warning against the sin of pride. But geological and archaeological evidence confirms that, a thousand years before Plato told the tale, a rich and powerful island civilization was indeed obliterated by a natural catastrophe, in the course of which the sea rose up and swallowed it whole.

The Aegean island-volcano of Thera, located in the Santorini archipelago, has erupted several times in its violent history, most spectacularly some 3,600 years ago in an event now known as the Minoan eruption. At the time of the eruption the island was the seat of a Bronze Age trading empire with strong cultural links to the Minoan civilization that flourished on nearby Crete. Recent archaeology has begun to uncover the splendours of Thera's lost city, Akrotiri, which for the past 3,500 years has lain buried beneath 30 m (100 ft) of volcanic pumice and ash. This Pompeii of the southern Aegean reveals Bronze Age Therans to have been a wealthy and sophisticated people: literate, artistic and intensely mercantile, their island an important trading hub between Europe, Africa and Asia.

But it was also a geological time bomb waiting to go off. Research led by the volcanologist Haraldur Sigurdsson in 2006

revealed the Minoan eruption to have been the second largest volcanic event in human history, an earth-shaking explosion that ejected some 60 cubic km (14 cubic miles) of volcanic material – ten times more than Vesuvius in AD 79 – and sent a series of powerful tsunamis racing across the southern Aegean which overwhelmed everything in their path.[2]

The eruption itself is likely to have been preceded by severe earthquakes across the region, a detail to which Plato alluded in his account:

> Now in this island of Atlantis there was a great and wonderful empire which had rule over the whole island and several others . . . But afterwards there occurred violent earthquakes and floods; and in a single day and night of rain all your warlike men in a body sank into the earth, and the island of Atlantis in like manner, and was sunk beneath the sea. And that is the reason why the sea in those parts is impassable and impenetrable, because there is such a quantity of shallow mud in the way; and this was caused by the subsidence of the island.[3]

The volcanic island of Thera erupted in c. 1620 BC, sending powerful tsunamis racing out across the southern Aegean. The destruction of the island – the seat of a wealthy Bronze Age trading empire – is thought to be the origin of the Atlantis story. This hand-coloured engraving shows the remains of the island in eruption once again in 1866.

In fact, all the details mentioned in Plato's account ring geologically true. A catastrophic 'rain' of ash and pumice did bury the entire island, much of which collapsed into an undersea caldera, while the 'impassable' barrier of shallow mud that blocked the surrounding sea routes is likely to have been an extensive raft of floating pumice, a maritime obstacle that can persist for many months in the wake of an island eruption. The collapse of the towering eruption column would have seen an avalanche of superheated gas and pumice surging into the sea, generating powerful tsunamis (Plato's 'floods') that then smashed into northern Crete, 110 km (68 miles) to the south, laying waste to coastal settlements. Thousands of people across Crete and other neighbouring islands are likely to have been killed by the associated earthquakes and tsunamis, as well as by the massive pyroclastic flows, and it appears that Minoan society never fully recovered from the impact. Over the course of the following century their culture declined and their surviving settlements were taken over by Mycenaean Greeks from the north. The identification of post-tsunami pottery on Crete –

'Insula Atlantis' from Athanasius Kircher, *Mundus Subterraneus* (1665). In Kircher's version, ancient Atlantis fills much of the north Atlantic, but all that is left now, he writes, are the Canary Islands and the Azores.

A late Minoan 'marine style' vase, decorated with writhing sea-creatures; a cultural memory of the inundation of Crete?

decorated in the so-called 'marine style' – gives a hint of the trauma suffered by the last of the Minoans: giant octopuses and other sea monsters writhe around the vessels, eerie memorials to the worst natural catastrophe to befall the ancient world.

Ancient tsunamis

The Atlantis story was not written down until a thousand years after the event it purportedly describes. The earliest tsunami for which there is contemporary documentation occurred in 479 BC at Potidaea, northern Greece, a Corinthian colony then under siege by the Persians. At a crucial point in the conflict, according to Herodotus, who recorded the event in Book VIII of his *Histories* (*c.* 440 BC), the Persian attackers attempted to exploit an unusual withdrawal of the sea around the peninsula, only to be caught out by the water's sudden return:

> When Artabazus had besieged Potidaea for three months, there was a great ebb-tide in the sea, lasting for a long while, and when the foreigners saw that the sea was turned to a marsh, they made to pass over it into Pallene. But when

they had made their way over two fifths of it and three yet remained to cross ere they could be in Pallene, there came a great flood-tide, higher, as the people of the place say, than any one of the many that had been before; and some of them who knew not how to swim were drowned, and those that knew were slain by the Potidaeans, who came among them in boats.[4]

Herodotus ascribed the cause of the sea's dramatic withdrawal and incursion to Poseidon's anger at the Persian attack on the temple at Potidaea. Perhaps the sea god's fury can be read as an analogy for the undersea earthquake that is likely to have caused the tsunami, for a tsunami it almost certainly was: recent excavations in the area have uncovered sediments laden with ancient seashells transported violently from the ocean floor, while computer models based on bathymetric data suggest that earthquakes and landslides in the region, combined with a colossal, bath-shaped depression in the sea floor off northern Greece, are easily capable of producing damaging tsunamis from

The Parting of the Red Sea in the film *The Ten Commandments* (1956), starring Charlton Heston.

15

The tsunami from Thera destroys a Minoan city on Crete, *c.* 1620 BC, in an illustration by Roger Payne for the children's magazine *Look and Learn.*

2 to 5 m (7 to 16 ft) in height. In the words of Klaus Reicherter, the palaeoseismologist in charge of the research, 'We found evidence of a tsunamigenic layer, dated with shells to 2500 BP, which may tentatively be interpreted as the sedimentary remains of the "Herodotus tsunami" of 479 BC.'[5]

Could a similar event have been the origin of the biblical story of the parting of the Red Sea? According to Exodus 14:21, 'the Lord caused the sea to go back by a strong east wind all that night, and made the sea dry land, and the waters were divided', allowing the Israelites to pass into safety, before the

sea returned and destroyed the pursuing Egyptians. Over the years researchers have examined a range of possible scenarios, including a tsunami generated by a local earthquake; but though the tsunami explanation remains 'technically feasible' (and accords with Moses' description of the waters standing 'upright as a heap'), the physical process that best fits the legend is a phenomenon known as 'wind setdown', which is essentially the opposite of a storm surge.[6] Wind setdown occurs when a strong wind, in combination with the prevailing tide, drives water abnormally far from the coast. By relocating the episode from the Red Sea to the Nile Delta – a plausible exchange, given a number of internal story details – the creation of a temporary wind-driven land bridge has been found to be meteorologically possible. Whether or not the event really took place is, of course, another matter.

The ancients were evidently familiar with tsunamis but they had no idea what caused them. Only the historian Thucydides, in his account of the Malian Gulf tsunami that struck in 426 BC during the Peloponnesian invasion of Attica, came close to correlating seismic and hydrographic events:

> About the same time that these earthquakes were so common, the sea at Orobiae, in Euboea, retiring from the then line of coast, returned in a huge wave and invaded a great part of the town, and retreated leaving some of it still under water; so that what was once land is now sea; such of the inhabitants perishing as could not run up to the higher ground in time. A similar inundation also occurred

Il mare! Il mare! Scene di indescrivibile panico si sono svolte in alcune città costiere del Giappone quando, in seguito ad un terremoto sottomarino, un'enorme muraglia d'acqua si è sollevata dal mare e si è abbattuta sulle strade travolgendo gli uomini come fuscelli, demolendo le case e provocando centinaia di vittime. (Disegno di W. Molini)

On average, Japan is hit by a tsunami every six and a half years. This newspaper illustration depicts a deadly tsunami striking Sagami Bay on 1 September 1923, following the Great Kanto Earthquake.

The tsunami of 1 September 1923 crashes into Yokohama, the second-largest city in Japan, causing severe damage. Illustration from *Le Petit Journal*, 16 September 1923.

at Atalanta, the island off the Opuntian-Locrian coast, carrying away part of the Athenian fort and wrecking one of two ships which were drawn up on the beach.[7]

Thucydides went on to consider the reason for such sudden risings of the sea. In contrast to Herodotus, who had credited Poseidon's oceanic temper, he ventured a more earthly explanation:

The cause, in my opinion, of this phenomenon must be sought in the earthquake. At the point where its shock has been the most violent, the sea is driven back and, suddenly recoiling with redoubled force, causes the inundation. Without an earthquake I do not see how such an accident could happen.

Thucydides was on the right track, but as will be seen in the following chapter, it would be many centuries before the causes of tsunamis were identified securely, at least in the seismically stable West, where such phenomena remain relatively rare.

In Japan, by contrast, tsunamis have always been a fact of life, 'a kind of geo-religious ritual', as Gretel Ehrlich describes them in her fine book *Facing the Wave* (2013).[8] More tsunamis strike Japan than any other country in the world – one every six and a half years, on average – with nearly 200 significant Japanese tsunamis recorded over the last 1,300 years. So it is no coincidence that 'tsunami' is a Japanese word (it translates as 'harbour wave', from *tsu* meaning harbour, and *nami* meaning wave, testament to the fact that these low-amplitude phenomena tend to pass unnoticed through deep water, only becoming apparent as they approach land).[9]

Detailed documentation is lacking about many of Japan's earliest tsunamis (though modern palaeoseismology is beginning to reveal much about their nature and extent), but one of the first to be officially recorded occurred in July 869, in the Sendai region of northeast Japan – the most tsunami-prone coastline on the planet. According to a contemporary chronicle, the *Nihon Sandai Jitsuroku* ('The True History of Three Reigns of Japan', completed in 901), the wave was preceded by a powerful earthquake:

On the 26th of the 5th month a large earthquake occurred in the province of Mutsu. The sky was illuminated like day-time. A little later, people, panic-stricken by the violent trembling, were lying on the ground; some were buried under fallen houses and others inside wide-opened ground fissures, while horses and cows desperately ran about and trampled each other . . . then roaring like thunder was heard toward the sea. The sea soon rushed into the villages and towns, overwhelming a few hundred miles of land along the coast. There was scarcely any time for escape, though there were boats and the high ground just before them. In this way about 1,000 people were killed. Hundreds of hamlets and villages were left in ruins.[10]

The 869 Jogan Sanriku earthquake, as it has become known, was an undersea quake with an estimated magnitude of 8.6 which occurred along the same section of the Japan Trench as the March 2011 event and, as was the case in 2011, the powerful tsunami it generated careered inland for around 4 km (2.5 miles), sweeping away all before it. The similarities between the two tsunamis led to many comparisons being made in Japan following the 2011 disaster, with renewed scientific attention being paid to the earlier event. In fact, in 2009, seismologists at Tokyo's National Institute of Advanced Industrial Science and Technology had argued that new evidence of the severity of the 869 tsunami should be applied to improving nuclear safeguards at the many reactors along the 'tsunami coast', where sedimentary records have revealed that a megatsunami strikes around once every 500 years. As the scientists pointed out, a major tsunami was already around a century overdue, but, as is so often the case, their warning was not taken seriously.[11]

Modern tsunamis

The March 2011 tsunami claimed nearly 16,000 lives and profoundly affected Japan's self-image as a disaster-ready society. Many of the victims along the Sendai coast would have had just enough time to escape – warning sirens went off within three minutes of the magnitude 9.0 earthquake, and in places the tsunami took an hour to arrive – but most had decided not to evacuate because they felt safe behind the large sea walls that had been built to protect them from tsunamis. Only two days earlier, on 9 March, a magnitude 7.2 offshore earthquake had shaken the same area of northeast Japan, generating a small tsunami that washed harmlessly against the coast's concrete perimeter. Two days later came another earthquake and another tsunami, but this time, as hours of horrifying camera footage show, those same sea defences were completely overwhelmed by the vast body of water that bulldozed its way inland. Japan was well prepared for tsunamis, but not for a tsunami as destructive as this, the result of the strongest earthquake in Japanese history and the fifth most powerful ever recorded on earth. The massive quake caused much of Japan's northeast coast to subside by

Emergency vehicles patrol the ruins of Sukuiso, Japan, following the tsunami of 11 March 2011.

nearly a metre, lowering the sea walls and allowing the tsunami to travel great distances inland. Its giant waves, measuring almost 39 m (128 ft), ravaged much of the coastline north of Tokyo, crippling the Fukushima Daiichi nuclear power station in the process, and precipitating Japan into a state of emergency not seen since the Second World War.

'3/11', as the event has become known in the Japanese media, exposed a dangerous flaw in the nation's otherwise impressive defences. The fact that the earthquake itself had been responsible

A year after the 2011 tsunami, untouched wreckage is still strewn over the abandoned coastal town of Namie, Fukushima Prefecture, evacuated when the nearby nuclear power station went into meltdown.

for only a handful of deaths was testament to the effectiveness of Japan's highly evolved building designs and its programmes of regular safety drills, but none of these antiseismic measures offered protection against the subsequent tsunami. The power station at Fukushima, for example, was virtually earthquake-proof, designed to shut itself down in the event of a major tremor. But the backup generators – there to power the station's emergency cooling system – had been installed at ground level or lower, with their fuel tanks housed outside behind a 5.7 m (18 ft) sea wall, a design that left the entire complex vulnerable to a bigger-than-average tsunami. Sure enough, the second of that afternoon's 10-m (32-ft) waves swept over the sea wall and destroyed the emergency fuel tanks, sealing the fate of the station. Over the following weeks three of Fukushima's overheating reactors went into meltdown, precip-itating the worst nuclear disaster since Chernobyl. As radiation levels almost a thousand times higher than normal built up inside the plant, the government imposed a 30-km (18-mile) exclusion zone around it, from which more than 200,000 people were evacuated, many of whom have yet to return home. Given the decades it will take to decontaminate the area, it is likely that most of them never will.

The catastrophe raised many questions about the future of disaster preparedness in Japan. The fact that the country's 16,000 km (10,000 miles) of concrete sea wall proved all but useless in the face of the tsunami was profoundly shocking to those coastal communities that had considered themselves pro-tected. Survivors in the worst-affected areas are still deciding whether to rebuild their tsunami-ravaged towns behind higher defensive walls, or to move away from the coastline altogether. Such an incursion, after all, was not without precedent. The Hokkaido tsunami of 1993 saw waves as high as a ten-storey building wash over the sea defences of northwest Japan. The port of Aonae had been completely surrounded by a solid tsunami wall, but it proved no defence against the mountainous waves that washed away most of the town.

But how high would a wall need to be to guarantee protec-tion from tsunamis? As we have seen, the maximum wave height

A fishing-boat lies grounded in the ruins of a port town on the southwest coast of Hokkaido, 14 July 1993, the day after a powerful tsunami struck, for which the sea defences were no match.

recorded in 2011 was an astonishing 38.9 m (128 ft): building a continuous sea wall of anything like that height would be economically impossible even were it technically feasible.

It is not just Japan that must live with such decisions; every inhabited coastline on the planet, from Scandinavia to Antarctica, has been struck by tsunamis in the past and remains vulnerable to them today. Indeed, as the rest of this chapter will show, some of the deadliest tsunamis in history have occurred in apparently unlikely locations, with the element of surprise serving only to compound their shocking capacity for destruction.

Lisbon 1755

The tsunami that followed the notorious Lisbon earthquake of 1 November 1755 took its victims completely by surprise. The powerful undersea quake (estimated magnitude 8.7) had brought down entire streets of houses and numerous churches that were packed with worshippers, it being the morning of All Saints Day, one of the holiest days in the Portuguese year. In the space of ten

A time-compressed view of the earthquake, fire and tsunami that destroyed most of Lisbon's city centre, at a cost of between 30,000 and 90,000 lives. Many of those who survived the earthquake escaped to the open spaces near the riverside, only to be engulfed by the tsunami that followed. This hand-coloured print dates from 1887.

minutes, much of the once-gilded city was reduced to ruins, beneath which many thousands of people lay dead or dying. As fires broke out amid the wreckage, crowds of survivors began to make their way through the rubble-strewn streets towards the open spaces along the banks of the river Tagus, the wide stretch of water that connects Lisbon to the rest of the world.

An hour after the earthquake, however, the water in Lisbon harbour suddenly receded, exposing a shipwreck-studded sandbank at the mouth of the estuary. The people who had sought refuge from the earthquake by clambering onto moored boats were the first to notice the drop in sea level, just before they also became the first to notice the 12-m (40-ft) wave rapidly advancing towards them from the west. One eyewitness, a British merchant named Daniel Braddock, wrote a vivid description of the mountain of water that ploughed its way towards the city:

> On a sudden I heard a general outcry, 'The sea is coming in; we shall all be lost.' Upon this, turning my eyes towards the river, which in that place is near four miles broad, I could perceive it heaving and swelling in a most unaccountable manner, for no wind was stirring; in an instant there appeared, at some small distance, a large body of water, rising as it were like a mountain, it came on foaming and roaring, and rushed towards the shore with such impetuosity, that we all immediately ran for our lives, as fast as possible; many were actually swept away, and the rest above their waist in water at a good distance from the banks.[12]

As Braddock and many other survivors observed, the tsunami was made up of three or more distinct waves, each of which crashed over the city's docks and into the riverside squares before washing back towards the estuary, dragging people and debris in their wake. In some places the waves advanced nearly a kilometre (around half a mile) inland, sweeping tonnes of wreckage along with them. Hundreds of survivors of the earthquake and fires who had congregated near the river for safety became victims of the unexpected tsunami.

It was not only Lisbon that suffered that day – the tsunami swept south over much of the Iberian peninsula, where it damaged the great sea wall at Cadiz, and on to the coast of Morocco. It also headed north to Britain, Ireland and the Netherlands, though the waves were much reduced by the time they reached the North Sea, being just big enough to damage a few barges in the lower Thames estuary; but it swept westwards with greater power, reaching the Caribbean some nine hours after the initial earthquake. At Antigua, according to an eyewitness, the sea level rose and fell 'twelve feet perpendicular several times, and returned almost immediately', while on other islands the waves rushed in, 'overflowing the low land; the people at Barbados were never more astonished; the rising water in Carlisle Bay appearing as black as ink instead of the clear sea-green'.[13]

The Lisbon earthquake and tsunami, the effects of which were felt across nearly a third of the planet, was, arguably, the first modern disaster, prompting the first international relief effort and establishing the now familiar protocols of emergency humanitarian response. It was also the first natural disaster to be studied scientifically and, as will be seen in the following chapter, it gave rise to the new science of seismology as well as to the world's first antiseismic building designs. It also established the long-term goal of predicting and warning against future disasters. This, of course, involves looking at the evidence of past events, and one of the more surprising seismological revelations to emerge in the wake of the Lisbon disaster was that the episode was far from unprecedented. We still tend to refer to 'the' Lisbon earthquake, while stressing how unexpected it was, but the area has in fact experienced major tsunamigenic quakes many times in the past: datable evidence on the ground points to significant precursors in around 5500 BC, 3600 BC, 60 BC and AD 382; while historical records confirm that a particularly destructive earthquake in January 1531 was also tsunamigenic, eyewitnesses having seen 'the Tagus river opened by its middle, splitting its waters into a pathway and showing the sand bed'.[14] Lisbon, it transpires, has always been in the firing line, and the only unusual feature of the 1755 event was its large death toll, reflecting the fact that the city's population had expanded since the previous earthquake, as it has continued to do ever since. The bad news is that Lisbon, inevitably, will go on to experience another tsunamigenic earthquake and, given the absence of any kind of disaster preparedness across the region, the death toll is likely to be high.

Krakatoa 1883

Civilization exists by geological consent, subject to change without notice: a lesson that was powerfully underlined in August 1883, when the eruption of Krakatoa destroyed many of the new ports and settlements that had been built by the Dutch colonial authorities around Indonesia's Sunda Strait, a strategically important shipping route between the islands of Java and Sumatra. Local legends told of fire-mountains and sea-ghosts summoning vengeful waves, but Dutch surveyors

The explosion of Krakatoa on 27 August 1883, from the popular children's magazine *Look and Learn*.

The paddle-ship
Berouw (from the
Dutch for 'remorse')
is carried inland by
the first giant wave
generated by the
eruption of Krakatoa.

The stranded
Berouw was carried
3 km inland.

had declared the local volcano, Krakatoa, to be extinct, and so the port towns of Anjer and Telok Betong were built in plain sight of the slumbering island. Krakatoa was, in fact, the active remnant of a megaeruption that had taken place some 1,500 years before. According to the Javanese *Book of Kings*, a rhapsodic history compiled by a nineteenth-century court poet, there had been an ancient eruption of 'the mountain Kapi', during which 'the whole world was greatly shaken, and violent thundering, accompanied by heavy rain and storms took place'. Afterwards, the sea rose and inundated the land, drowning the inhabitants of the northern part of the Sunda country: 'After the water subsided the mountain Kapi and the surrounding land became sea and the Island of Java divided into two parts.'[15]

Though it had rumbled intermittently over the centuries, Krakatoa remained relatively quiet until May 1883, when a series of explosions announced the fact that the island volcano had awoken. Its towering ash cloud, rising many kilometres into the stratosphere, became a short-lived tourist attraction before the eruption appeared to quieten down in June. Then, on 27 August, entirely unexpectedly, Krakatoa blew itself to pieces. The explosion – which created the loudest sound ever heard on earth – jettisoned 46 cubic km (11 cubic miles) of fragmented rock, ash and gas, and sent a series of powerful tsunami waves crashing into the surrounding shorelines.

The first and second waves hit the Sumatran town of Telok Betong, tearing ships from the harbour and hurling them ashore. One of these vessels, the armed paddle steamer *Berouw* (Dutch for 'remorse') had been left stranded on her side in the middle of the town, her entire 28-man crew killed on impact. What happened next would become a lasting image of the terrible power of tsunamis: the third and final wave, a monstrous wall of black, debris-laden water more than 40 m (130 ft) high, reared out over the Sunda Strait, obliterating everything in its path. Not a building was left standing in the Javan port of Anjer, or in any of the other 300 towns and villages swept aside by the wave. On Sumatra, the beached *Berouw* was picked up and carried 3 km (almost 2 miles) inland, before finally being deposited in a jungle

An ocean of dead, from
Camille Flammarion's
La Fin du Monde (1893).

valley some 18 m (60 ft) above sea level. But the town it had left
behind had been erased entirely, leaving what a later eyewitness
described as 'a plain bare and laid waste. Nothing is left of Telok
Betong or any of the surrounding villages.'[16]

At least 36,000 people were killed by the giant waves, many
of them swept out to sea where they formed a grisly obstacle to
shipping, as the crewman of a British merchant ship recalled:
'masses of dead bodies, hundreds and hundreds striking the ship
on both sides – groups of 50 to 100 all packed together, most of

them naked'.[17] Even a tiger was seen among the dead, testament to the tsunami's ferocious power.

The wave, meanwhile, headed towards the southeast entrance of the Sunda Strait, from where it radiated out across the Indian Ocean, gaining speed as it found deeper water. Some five and a half hours later the tsunami reached the east coasts of India and Sri Lanka: 'the wave reached halfway up to Calcutta on the Hooghly', as the Tidal Survey of India recorded, while the *Ceylon Observer* carried a report that in the southern port of Galle the sea had receded to the end of the jetty before rushing back in with surprising force. A few kilometres up the coast, a woman working in a paddy field was swept away and drowned, the most distant casualty of the eruption. Just as in December 2004, the majority of those killed by the tsunami died in Indonesia, but its power and endurance was brought home by deaths in Sri Lanka and elsewhere, as well as by the astonishing distances travelled by the waves: some ten hours after the eruption 'a great tidal disturbance' was reported at Aden on the Arabian peninsula, while metre-high waves were also seen to strike the harbours at Cape Town and Port Elizabeth on the South African coast, more than 7,500 km (4,500 miles) away from the now dematerialized volcano.[18]

Even though Krakatoa had been a small volcanic island in a remote patch of ocean, the world had soon come to know of the eruption, for by 1883 there were thousands of kilometres of submarine telegraph cable connecting every continent on earth. The urgent brevity of the telegraphers' messages had given rise to a new kind of truncated poetry that was well suited to conveying the horrors of a natural disaster. Thus, the scene around the Sunda Strait the day after the explosion was haunt - ingly summarized in a five-word telegram sent by the Lloyd's shipping agent in Serang: 'All gone. Plenty lives lost.'[19]

Sanriku 1896

One of the deadliest tsunamis in Japanese history occurred on the evening of 15 June 1896, after a 'slow' undersea earthquake

A numbed survivor amid the ruins of Karakuwa village, Miyagi Prefecture, in the aftermath of the 1896 Meiji Sanriku tsunami, in which 27,000 people died.

ruptured a section of sea floor along the Japan Trench around 160 km (100 miles) off the northeast coast of Honshu. That summer Sunday had been a day of national celebration to honour soldiers returned from the Sino-Japanese War, so the beaches were crowded with holidaymakers, while wedding parties and civic celebrations were still in full flow as evening began to fall. When the tremors were felt, at around 7.30 p.m., they were deceptively small. Just another everyday earthquake, it was thought, and so the festivities continued. But in fact a massive tsunami had been generated by a sudden sea-floor slippage. The first wave arrived about half an hour later, a devastating wall of water that crashed into the Sanriku coast with a noise like a raging hailstorm, a second wave following close behind. The death toll was enormous, and in many coastal settlements the only survivors were fishermen who had been at sea when the earthquake struck. Out in deep water, the tsunami swell would have passed under their boats in the form of an imperceptible high-speed ripple, and it was only when they returned home in the morning that the men discovered the carnage.

So many people were dead or missing that the authorities gave up counting the victims and counted the survivors instead, while a pall of 'sad smoking fires' hung over the Sanriku coast from the burning of thousands of bodies, many of which had been too badly damaged to be identified.[20] More than 27,000 people died and dozens of fishing villages were entirely destroyed, but the Sanriku coast was not abandoned by the survivors. Over the years the towns and villages were all rebuilt, albeit a little higher and further from the sea, behind some of those ancient tsunami stones that still dot the Japanese coast.

Hilo 1946

As will be seen in the following chapter, it was the Sanriku disaster that first alerted the Anglophone world to the Japanese word 'tsunami', but it was an event that took place 50 years later that made it a household word: the April Fool's Day tsunami of 1946, which devastated the Hawaiian town of Hilo (pronounced 'he-low').

The tsunami had its origins more than 3,700 km (2,300 miles) away, on the northern slope of the Aleutian Trench, a 3,200-km-long subduction zone that curves along the Alaska peninsula. Here, the northern edge of the Pacific plate creeps beneath a section of the North American plate at a rate of nearly 4 cm (1.5 in.) per year, causing regular earthquakes and volcanic eruptions along the length of the Aleutian Island chain. As is the case along the Japan Trench, the majority of these near-continuous tremors cause little in the way of damage or disruption, but the minute-long earthquake that occurred at 2.29 a.m. (local time) on 1 April 1946 was fated to be different. Measuring 7.8 on the recently devised Richter scale, its vibrations triggered an underwater landslide several kilometres east of the earthquake's epicentre, which in turn generated a powerful tsunami that radiated through the surrounding water. Forty minutes later, the southern shore of Unimak Island was hit by a 30-metre (100-foot) wave of displaced water which smashed into the newly built Scotch Cap Lighthouse, sweeping it from the rocks and killing all five coastguards on duty.

The shock waves, meanwhile, continued south through the open ocean, picking up speed as they headed into deeper waters.

Running from the wave: this dramatic photograph, taken by a local barber, Cecilio Licos, shows the third wave of the tsunami train crashing into downtown Hilo on the morning of 1 April 1946.

The wave surges over Hilo's commercial pier, 1 April 1946. The stranded stevedore in the photo was one of the tsunami's 165 casualties.

Soon they were racing across the Pacific at nearly 800 km/h (500 mph) – as fast as a jet airliner – with their shallow crests around 100 km (60 miles) apart. The waves had reduced in height when they entered deeper waters, but even in the form of high-speed ripples they had lost none of their destructive potential.

Unusually, the first wave had been detected while it was still out at sea. The radio operator on the USS *Thomson*, which was approaching Pearl Harbor in the early hours of 1 April, had picked up a message from a passing patrol plane, reporting an unusual movement on the surface. But when the pilot was ordered to investigate, whatever it was had already disappeared, having apparently outpaced the speeding aircraft; a claim that the crew of the *Thomson* dismissed as an April Fool. But, not long after, a message was received that the naval fleet at Pearl Harbor had just been knocked about by an unexplained tidal surge, and the *Thomson* was ordered to stay clear of the harbour until the cause of the disturbance was known.

Almost inevitably, given the date of the disaster, the only warning that was given out was ignored by everyone who heard it. A well-known breakfast radio presenter broadcasting from Oahu – which had already been hit by the first of the tsunami waves at 6.30 a.m. – interrupted his programme to issue a wave warning to the rest of the Hawaiian islands. But one of the best-loved features of the presenter's show was the quality and inventiveness of his April Fools, as a consequence of which, until the waves came crashing into the Big Island coast 25 minutes after the announcement was made, no one who heard it had believed it.

But it was true, and the trio of waves brought death and devastation to Hilo Bay, a palm-fringed curve of coastline lying directly in the firing line south of the Aleutian Trench. A total of 159 people were killed across Hawaii, 24 of them at Laupahoehoe School, where many of the teachers and pupils had wandered onto the exposed sea floor to marvel at the unexpected rock pools. 'The ocean sucked out like a bathtub emptying', one of the surviving teachers, 21-year-old Marsue McGinnis, later described, with the first of the waves arriving soon after. This first wave was strong but not particularly dangerous, she recalled: 'it came up a little bit above the high-water mark. So we looked at that, "That's a tidal wave? Something's wrong here, you know."'[21] No wonder some of the children asked her why it was called a 'tiger wave'.

A second wave came, which was a little bigger than the first, washing away a couple of boat sheds, but the third wave, which arrived about twenty minutes later, was something altogether different, a fast-moving wall of grey-black water advancing towards the shore 'with a roar like all the winds in the world', as McGinnis recalled. The immense wave, which powered inland at more than 100 km/h (60 mph), wrenched 8-tonne blocks from Hilo's stone breakwater and hurled them into the bay-front streets, smashing through the wood-framed buildings like a naval cannonade. Many of the buildings on the seaward side were pushed 20 m (60 ft) across the road and crushed against the buildings on the opposite side, while most of the coastal region's

A railroad car shoved under a bakery building by the power of the wave. Hilo, April 1946.

bridges, railway lines, roads and docks were also put out of action. It was an island-wide disaster, made worse by its being entirely unexpected. But Hawaii has a long history of tsunami inundation that is reflected in its language and its folklore, so why should this event have come as such a surprise?

It was partly due to the fact that the originating earthquake took place thousands of kilometres away, so no tremor was felt beforehand, and partly because it had been more than twenty years since the last big tsunami had hit the islands, in 1923, time enough for people to forget how to read the warning signs. Worse, a local tsunami warning system that had proved effective against a small but powerful inundation in 1933 had recently been abandoned due to a high number of inconvenient 'false alarms'; and anyway, there were no night staff at the Hawaiian observatory to have monitored the faraway earthquake. By the time the seismologists were setting off for work at 7 a.m., the waves were already crashing into the coast.

This was the event that led to the establishment of the Pacific Tsunami Warning System, an impressive combination of technology and surveillance that has contributed to the safety of tsunami-prone communities for nearly 60 years. But, as the next two examples show, even the most effective warning systems are limited by how much we really know about our planet's restless behaviour, and by its almost infinite capacity for surprise.

Papua New Guinea 1998

One of the twentieth century's most destructive tsunamis hit the northwest coast of Papua New Guinea on the evening of 17 July 1998. The tsunami was generated by a massive underwater landslide that had in turn been precipitated by a relatively modest (magnitude 7.0) offshore earthquake along the boundary of the Australia and Pacific plates. Though the earthquake was registered on the region's many seismographs, its small size provoked no fears of a tsunami. Indeed, the Pacific Tsunami Warning Center in Hawaii issued an immediate advisory bulletin that read: 'THIS IS A TSUNAMI INFORMATION

Papua New Guinea: more than 2,200 people were killed by the tsunami of 17 July 1998. Many of the dead were buried where they were found, in makeshift graves dug in the sand and covered over with driftwood.

MESSAGE, NO ACTION REQUIRED ... NO DESTRUCTIVE PACIFIC-WIDE TSUNAMI THREAT EXISTS.'[22] So when the train of three 10 to 15-m (30 to 60-ft) waves slammed into a 30-km (19-mile) strip of the island's isolated northwest coast, they did so completely by surprise. For while it was true that no Pacific-wide tsunami had been generated by the shallow-focus earthquake, the ensuing landslide created a powerful local tsunami that entirely destroyed a number of coastal villages and claimed more than 2,200 lives, including many children home from their mission school holidays. Bodies continued to be found weeks later, buried in sand and mud, or washed up onto the shores of Indonesia, 160 km (100 miles) to the west; while many of the survivors were left badly injured and vulnerable to infection, having been thrown against trees, impaled on mangrove stumps or lacerated by bacteria-laden coral. As one of the visiting doctors later observed, the scene was like the aftermath of a battle.

The 1998 tsunami proved disturbing for seismologists, for it revealed a hitherto unknown potential for small earthquakes to trigger large tsunamis. Such a chain of events is particularly

41

dangerous, as the originating earthquake may be too weak to be felt on land, and too minor to prompt a tsunami warning. Any resulting waves will therefore appear unannounced, apart from the tell-tale withdrawal of the sea – though significant with-drawals are not a feature of every tsunami, as was the case in 1998. More worryingly, as the geologist Hugh Davies discovered when he interviewed survivors on the island, there is a widespread lack of tsunami awareness across Papua New Guinea, especially among the young, for whom traditional social and storytelling bonds are less important than they were for their parents. 'Communal memory is short', he notes, and though a tsunami strikes somewhere on the islands on average every 15–70 years, this is ample time for a population to forget how to read the signs.[23] So in spite of our technology and our best intentions, there will always be times when the oceans attack without warning; a lesson that was brought home on a larger and more shocking scale on the morning of Boxing Day 2004.

Indian Ocean 2004

Just before eight o'clock in the morning of 26 December 2004, the third-largest earthquake ever recorded struck 160 km (100 miles) off the northwest coast of Sumatra. About 30 km (19 miles) below the seabed, a 1,200-km (750-mile) section of fault line between the Indian Plate and the Burma Plate suddenly slipped by as much as 15 m (nearly 50 ft). As a consequence, the seabed lurched upwards, displacing millions of tonnes of water above it; as the uplifted area collapsed, the water gushed away in all directions in the form of a powerful tsunami.

   The earthquake itself had been severe, and was felt over much of South East Asia – indeed, it caused the entire planet to vibrate, shortening the day by three-millionths of a second – although the only structural damage occurred in the immediate vicinity of northwest Sumatra, where buildings in the town of Banda Aceh were badly shaken. But it would not be long be-fore the extent of the earthquake's oceanic impact became apparent. By 8.15 a.m. the tsunami had slammed into the

Acehnese coast, on the northern tip of Sumatra, surging inland and drowning more than 150,000 people. Fifteen minutes later it hit the Andaman and Nicobar islands, then, an hour on from that, the waves made landfall along the already crowded beaches of southern Thailand.

All this time, the tsunami was also moving westwards across the Indian Ocean, where in two hours it had reached the eastern shores of Sri Lanka, and in three hours it had crashed into the low-lying Maldives, washing over some of the islands entirely. But the tsunami was far from finished: seven hours after the quake and more than 4,800 km (*c.* 3,000 miles) away from the epicentre, the waves struck the coast of Somalia, on the opposite side of the Indian Ocean, where they killed at least 150 people and injured many more. In Tanzania, a number of swimmers were killed in the sea off Dar es Salaam, while the two confirmed deaths on the coast of South Africa, twelve hours on from the earthquake itself, were the day's most distant casualties.

So why was no tsunami warning given? The four-minute earthquake had registered on the world's seismographs even as it was still in progress, but the absence of sea-floor sensors in any of the affected areas meant that there was no way of confirming if a tsunami had been generated. Scientists on duty at the Tsunami Warning Center in Hawaii were well aware that a tsunami was likely to have been generated by the massive submarine rupture, but their warning system had been equipped to monitor only the coastlines of the Pacific. All they could do was email a general information bulletin to the 26 Pacific nations on their contact list, and wait for further details to emerge.

Just over an hour later, as the first wave was battering the coast of Thailand, staff at the centre emailed a second, upgraded warning bulletin to the contacts on their list, and had begun a hopeless search for the telephone numbers or email addresses of civil defence coordinators across the Indian Ocean region but, as Barry Hirshorn, one of the geophysicists on duty that day, explained in an interview with *Channel 4 News*, 'there were no contact points, no organizations, no warning systems that I knew of in the area' – and so, as the morning went on, the waves continued to plough into some of the world's most densely populated coastlines, with no warnings given out by the only people on the planet who could have known, at that moment, just how destructive the events of the following few hours would prove.

One of the many disheartening details that emerged in the wake of the disaster was that only the previous year governments from around the Indian Ocean had called a meeting to plan the installation of a tsunami warning system, only to vote against it on economic grounds. Since the last major tsunami in the region had occurred in 1833 (with a smaller one 50 years later, sent over from Krakatoa), the risk of another had seemed disproportionate to the cost of installing the equipment. But even had they agreed to act, how much difference could it have made? A year would not have been enough time either to install the system or, more crucially, to educate an entire coastal population in how to respond to alerts. As is shown by so many episodes in this book, the history of warning systems is not a particularly happy one, beset as it is by human and technological errors. And though we

This solidly built mosque is all that remains of a coastal town in Aceh province, Indonesia, in the wake of the 2004 Boxing Day tsunami. The disaster claimed more than 230,000 lives in fourteen countries across two continents.

have discovered a great deal about tsunamis over the years, there remains a great deal more that we haven't. The following chapter will look at the history of tsunami science, before outlining what we currently know, and what else we urgently need to learn before the next tsunami strikes.

2 The Science of Tsunamis

'Well,' I thought, 'you're a pretty poor oceanographer not
to know that tsunamis increase in size with each new wave.'
As soon as possible I began to look over the literature, and
I felt a little better when I could not find any information
to the effect that successive waves increase in size, and yet what
could be a more important point to remember?'[1]

On the afternoon of 20 February 1835, during his fourth year as
naturalist on board HMS *Beagle*, Charles Darwin was fossicking
in some woodland near the port of Valdivia, southern Chile,
when he felt what he guessed to be a powerful earthquake, the
motion of which 'made me almost giddy'.[2] The trees remained
unmoved by the two-minute tremor, he noted, but 'the tides
were very curiously affected', with several large waves sloshing
about on the nearby shore. But an earthquake in Chile is a
common occurrence, and the 25-year-old Darwin gave it no more
thought until two weeks later, when the *Beagle* arrived at
Concepción harbour, around 400 km (250 miles) north of
Valdivia, to find a scene of utter devastation, 'the whole coast
being strewed over with timber and furniture, as if a thousand
ships had been wrecked'. Walking along the wave-shattered
shore, Darwin observed that hundreds of pieces of undersea rock
had been cast high onto the beach: 'One of these was a slab six
feet by three, and about two feet thick.' The wave power needed
to move such objects was almost unimaginable.

As Darwin noted, the neighbouring port towns of
Concepción and Talcuhano had been all but destroyed by the
earthquake, though most of the damage to the latter had been
caused by a subsequent tsunami: 'a great wave, which, travelling
from seaward', burst over Talcahuano, sweeping away the remains
of the fallen buildings. Darwin's detailed description of the
tsunami (though 'tsunami' was not a word he used) was based on
local eyewitness accounts, and is a masterpiece of reportage:

Shortly after the shock, a great wave was seen from the distance of three or four miles, approaching in the middle of the bay with a smooth outline; but along the shore it tore up cottages and trees, as it swept onwards with irresistible force. At the head of the bay it broke in a fearful line of white breakers, which rushed up to a height of 23 vertical feet above the highest spring-tides. Their force must have been prodigious; for at the Fort a cannon with its carriage, estimated at four tons in weight, was moved 15 feet inwards. A schooner was left in the midst of the ruins, 200 yards from the beach. The first wave was followed by two others, which in their retreat carried away a vast wreck of floating objects. In one part of the bay, a ship was pitched high and dry on shore, was carried off, again driven on shore, and again carried off . . . The great wave must have travelled slowly, for the inhabitants of Talcahuano had time to run up the hills behind the town; and some sailors pulled out seaward, trusting successfully to their boat riding securely over the swell, if they could reach it before it broke. One old woman with a little boy, four or five years old, ran into a boat, but there was nobody to row it out: the boat was consequently dashed against an anchor and cut in twain; the old woman was drowned, but the child was picked up some hours afterwards clinging to the wreck. Pools of salt-water were still standing amidst the ruins of the houses, and children, making boats with old tables and chairs, appeared as happy as their parents were miserable.[3]

The aftermath that Darwin described is now tragically familiar. Television images of the 2004 Indian Ocean tsunami and the 2011 Japan tsunami conveyed many of the same unforgettable details that featured in Darwin's account: a huge implacable wave shoaling in off the horizon; its force and relentlessness as it surges ashore, picking up entire fishing fleets and hurling them inland; stories of loss and of miraculous survival; and the pain and resignation of the survivors. In the years since Darwin's encounter with the Chilean tsunami,

dozens of similar episodes have occurred in every ocean of the world. In fact, a tsunami will strike somewhere on earth almost every year. But how much do we really know about these catastrophic waves, and what is there still to be discovered?

By the time Darwin walked among the ruins of Talcuhano, the foundations of modern seismology had been well established. Eighty years earlier, in the wake of the Lisbon earthquake, eyewitnesses and survivors had been sent detailed questionnaires about the undersea tremor and its effects. Questions such as 'How long did the earthquake last?'; 'How many aftershocks were felt?'; 'Did you notice what happened to the sea, to fountains and to rivers?'; 'Did the sea rise or fall first, how many hands did it rise above normal, and how many times did you notice the extraordinary rise or fall?' yielded hundreds of responses which, when collated and cross-referenced by scientists across Europe, revealed a wealth of information about the earthquake and tsunami.[4] It was the first systematic attempt

Nothing is left standing in Arica, Chile, after a large earthquake and subsequent tsunami flattened the town in August 1868.

to create an objective overview of a natural disaster, and it proved to be a landmark moment in the development of seismology.

John Michell, a young Cambridge naturalist, made the first thorough synthesis of the Lisbon reports, and from them he drew some extraordinary conclusions. Michell's now-celebrated paper, *Conjectures concerning the Cause, and Observations upon the Phænomena, of Earthquakes* (1760), began with a summary of eyewitness material relating to historical disasters. From this, he observed that certain regions of the earth are 'subject to the returns of earthquakes', that is, they are more seismically active than others, which might seem obvious today, but at the time it represented a crucial advance in seismic understanding.[5] Earthquakes are not randomly distributed, but are clustered in areas of repeat activity, which meant that the task facing seismology was not guessing where the next quake might occur, but discovering what goes on in the strata beneath earthquake-prone regions.

Michell applied the latest geological knowledge to the puzzle of earthquake production, arriving at a novel hypothesis: subterranean steam, generated from seawater vaporized by molten rock, made its way between uniform strata until it reached a discontinuity near the surface. The steam then rushed towards that area, its movements sending powerful seismic waves hurtling through the earth's crust; at the surface these waves would be experienced as earthquakes. Michell, of course, was wrong about the steam – the theory of plate tectonics lay two centuries in the future – but he was right about the propagation of seismic waves through solids which, in an inspired analogy, he likened to a carpet being 'raised on one edge and then suddenly brought down again on to the floor', the shock waves being seen to pass along the entire length of fabric.[6]

The completed questionnaires from Lisbon had confirmed that the initial shock waves were experienced that morning at different times and at varying intensities, depending on location, while the interval between the earthquake and the arrival of 'the succeeding wave' (in other words, the tsunami) increased as one moved along the coast. Michell believed that it would be possible to identify the earthquake's source by measuring the waves' direction and arrival times from a series of different locations. Thus, using a range of seismic observations and tsunami arrival times, he estimated the earthquake's epicentre to have been 'under the ocean, somewhere between the latitudes of Lisbon and Oporto (though probably somewhat nearer to the former) and at the distance, perhaps, of ten or fifteen leagues from the coast', which, all things considered, was a pretty impressive estimate.[7]

Michell also applied himself to the mechanics of tsunamis, which he attributed to the sudden subsidence of the earth above a submarine epicentre: 'The waters will flow every way towards it, and cause a retreat of the sea on all the shores round about: then presently . . . the earth will be raised, and the waters over it will be made to flow every way, and produce a great wave immediately succeeding the previous retreat.'[8] Again, Michell's explanations were on the right track, but his most impressive insight was that tsunami waves travel at different speeds through varying depths of water:

This photograph, taken over Miyagi Prefecture on the afternoon of 11 March 2011, shows the tsunami surging inland for several kilometres.

It is observable that the times which the wave took up in travelling are not in the same proportion with the distances of the respective places from the supposed source of the motion ... the true reason of this disproportion, seems to be the difference in the depth of the water; for in every instance, the time will be found to be proportionably shorter or longer, as the water through which the wave passed was deeper or shallower.[9]

Michell had hit upon one of the key characteristics of tsunamis: their variable speeds. Tsunamis are as deep as the oceans, generated as they are by a disturbance of an entire water column, so in contrast to wind-driven surface waves, their great depth and long wavelengths means that hardly any wave energy is lost through friction – in fact, far from losing their impetus at sea, tsunami waves actually *gain* speed as the water deepens, flattening into a series of imperceptible ripples, sometimes hundreds of kilometres apart. As has been seen, these ripples can travel at enormous speeds, depending on the depths through which they pass. In the deep mid-ocean waters of the Pacific they can reach speeds of up to 800 km/h (500 mph) – as fast as a jet plane – while in shallower waters, such as Indonesia's Sunda Strait, their speed can be as low as 100 km/h (around 60 mph): still too fast, however, to outrun or outswim.

But it is only when they make landfall that we see their destructive potential. In open water the waves' low amplitudes cause only slight swells to form, but as they approach land they quickly lose speed as shelving coastlines and shallower waters apply friction to the base of the waves. As the waves slow down, they begin to rise, with the waves at the back catching up with the leaders, shunting together to form precipitous walls of water. Long-wavelength tsunamis can travel either peak first or trough first, depending on the particular mechanics of their production. The characteristic withdrawal of the ocean in advance of an incoming wave is actually a trough arriving ahead of the first peak, and is one of the clearest signs of an approaching wave. But unlike plunging breakers, which dissipate their energy as soon as they hit the surf zone, long tsunami waves will continue far inland, sometimes for many kilometres, with the entire weight of the ocean behind them. As many eyewitnesses have testified, they can seem more like floods than conventional waves, 'like a force pushing the water behind', as one survivor of the April Fools' Day disaster described it: 'It wasn't a rolling kind of wave that you could think of. I said, a couple of times, it's like filling a cup like this, just pour the water in, when you fill it up to the top it flows over.'[10] This overflowing is a consequence of a tsunami's long wavelength – the distance between the wave crests – and the longer the wavelength, the further inland each wave will flow until it eventually 'breaks' and begins its destructive retreat.

This pattern may well repeat itself several times in a classic tsunami wave-train, with up to an hour's gap between the waves: so, just when the survivors of the first wave think that the worst is over, an even bigger second wave follows, and then perhaps a third or a fourth, each one wreaking further destruction. It was this that caught out the marine geologist Francis Shepard, who in April 1946 was one of a small community of American scientists stationed on Hawaii, waiting to observe the atomic bomb tests at Bikini Atoll, 4,000 km to the west. Their presence would make the April Fool's Day disaster the most studied tsunami in history, until the morning of Boxing Day 2004.

Shepard was asleep in his rented cottage on Oahu's north shore, when he and his wife were woken at 6.30 a.m. by what he described as 'a loud hissing sound, which sounded for all the world as if dozens of locomotives were blowing off steam directly outside the house'.[11] This was the first of the wave train making landfall. Shepard grabbed his camera and ran from the cottage to investigate, witnessing a sharp withdrawal of the water, which left the reef entirely exposed to view. Although aware that another wave was likely to follow, Shepard had seriously underestimated its size and ferocity, and so he stayed on the beach, waiting to photograph the incoming sea. Suddenly he realized that the second wave massing on the distant horizon was far more powerful than the first. A couple of minutes later,

The first wave crashes over Coconut Island, Hilo Bay, 1 April 1946.

it had ploughed into the cottage – 'the refrigerator passed us on the left side moving upright out into the cane field', as he recalled – but by the time the third wave in the train arrived, destroying the beachside house and claiming the lives of six people further along the shore, Shepard and his wife had already made their way to higher ground.[12]

Having witnessed a tsunami at such close proximity, Shepard was determined to understand the phenomenon better, with a view to contributing to the development of some kind of coastal warning system. He had, after all, only just escaped with his life, but more than 150 others had not been so lucky. Nine months later, he and two co-authors published what proved to be an influential paper in the inaugural issue of the journal *Pacific Science*, which began with a useful linguistic evaluation of the range of available synonyms for 'tsunami':

Tsunamis are also sometimes termed 'seismic sea waves', and are popularly known as 'tidal waves'. The latter term is patently undesirable, as the waves have no connection whatsoever with the tides. 'Tsunami' is used herein in preference to 'seismic sea wave' because of its greater brevity, and because the etymological correctness of the term 'seismic sea wave' appears open to question.[13]

The 'tidal wave' misnomer presumably derived from the tide-like behaviour exhibited by tsunamis which, as we have seen, rarely resemble conventional waves, although a tsunami approaching shallow water can sometimes shoal into a bore-shaped wave with steep, step-like fronts. Bores, unlike tsunamis, are true tidal waves, resulting from large tides surging into rivers and estuaries, but though their causes are different, their effects on land can be similar. The notorious *mascaret*, for example, was a regular tidal bore that caused great damage and loss of life along the lower Seine in Normandy, although a century of dredging and development has now consigned it to the past. The largest remaining tidal bore occurs at the mouth of the Qiántáng Jiang river, near Hangzhou in China; it can reach heights of 9 m (30 ft) and is often extremely dangerous, as dozens of sight-seers discovered in August 2013 when the fast-moving bore washed them into the river.

Rogue waves, by contrast, are single, mountainous ocean waves that can appear as if from nowhere. Their cause is not fully understood, but they are thought to arise from the anomalous focusing effect of strong winds and powerful currents that serve to unite a series of regular waves into one vast wave that can rear up from the surrounding sea like a tsunami approaching land. Though rare, they present a serious hazard to shipping.

The term 'seismic sea wave', meanwhile, can be just as misleading as 'tidal wave', since not all tsunamis are generated by seismic activity: landslides, rockfalls and volcanic eruptions have all been known to create catastrophic waves, while tsunamis have, on rare occasions, been caused by meteorites or other extraplanetary objects, as depicted in the film *Deep Impact* (1998). There has even been the odd case of tsunamis resulting from gas hydrate eruptions or nuclear weapons tests; the vast majority of tsunamis, however, are generated by submarine earthquakes.

This image, from 1903, shows sightseers watching the *mascaret* (tidal bore) on the River Seine at Caudebec-en-Caux in Normandy. The bore resulted from large spring tides, and was the frequent cause of damage to ships, as well as the occasional fatality. After dredging and other precautionary measures the Seine bore has all but disappeared.

Le Petit Parisien

SUPPLÉMENT LITTÉRAIRE ILLUSTRÉ

TOUS LES JOURS
Le Petit Parisien
(six pages)
5 centimes

CHAQUE SEMAINE
LE SUPPLÉMENT LITTÉRAIRE
5 centimes

DIRECTION: 18, rue d'Enghien (10e). PARIS

ABONNEMENTS

PARIS ET DÉPARTEMENTS:
12 mois, 4 fr. 50. 6 mois, 2 fr. 25

UNION POSTALE :
12 mois, 5 fr. 50. 6 mois, 3 fr

LE MASCARET. — A CAUDEBEC-EN-CAUX

Causes of tsunamis

As Shepard's paper also pointed out, though most tsunamis are linked to earthquakes, the earthquakes themselves are rarely the direct cause, 'rather, both are caused by the same sudden crustal displacement'. This was an important insight, for though we know that the majority of tsunamis are associated with under-sea earthquakes, we do not always know the exact mechanism involved in generating the waves, whether sea-floor subsidence, a subsequent landslide or some other submarine process.

Not every kind of earthquake is capable of producing a tsunami. The 1906 San Francisco earthquake, for example, sent seismic shock waves far out to sea, but because the San Andreas fault is a lateral rather than a vertical slip, its vibrations rarely lead to a tsunami. For a tsunami to be generated, there needs to be a substantial vertical movement of an area of sea floor, which is usually associated with plate boundary slippages along major subduction zones. These are long seams in the earth's crust where one tectonic plate slides, or subducts, below another, melting

A rogue wave approaches a merchant ship in the Bay of Biscay, c. 1940. Rogue waves are believed to form from a series of waves joined together through a combination of strong winds and fast currents.

back into the hot mantle like the disappearing end of a conveyor belt. The process is rarely smooth, however, and sections of a subducting plate often stick to the stationary plate, pulling it down over a long period of time before the accumulated strain gives way and the released section jolts back into place.

This is what happened on Boxing Day 2004, when a long-stuck section of subducting plate along the Java Trench was suddenly released, causing a 1,200-km (750-mile) section of sea floor to 'jump' by several metres, sending an enormous volume of displaced seawater racing across the Indian Ocean in the form of powerful waves. This kind of colossal rupture, known as a megathrust earthquake, has been responsible for generating many of history's most lethal long-range tsunamis.

A tsunami's destructiveness depends on more than just the size and duration of the originating earthquake; distance from the coast plays an important part, as does the often complex influence of local coastal topography. The latter can make the deciding difference between life and death on shore. A stretch of coastline fronted by a steep seabed may escape a tsunami relatively unscathed, while a neighbouring area with a shallower seabed suffers the full effects of the incursion. In Thailand, local bathymetry and beach alignment were determining factors in the fates of many of those caught up in the Boxing Day tsunami. Karon Beach on the west coast of Phuket, for example, is fronted by a sand dune that did much to reduce the energy of the incoming wave, while Kamala Beach, a few kilometres to the north, has a large near-shore rock platform that allowed the wave to barrel inland, leading to significant loss of life. Khao Lak, too, suffered particularly badly due to a wide section of shallow sea floor that rises gently in front of the beach and allowed the wave to surge inland at full strength.

In Sri Lanka, too, the busy fishing village of Oluvil became the first settlement to be hit by the tsunami, but only two of the villagers died, due to the lucky effect of a nearby submarine canyon: its depth and shape had served to dissipate wave energy, thereby reducing the wave height. A few kilometres up the coast, however, the town of Kalmunai was almost completely

At the subduction zones, a dense oceanic plate dives (subducts) beneath a lighter continental plate, with strain building up when they get stuck. When the stuck plates lurch free, causing an earthquake, a large area of sea floor can jump, triggering an associated tsunami.

destroyed by the same wave, with the loss of more than 8,500 lives. Their misfortune was the presence of a nearby submarine ridge that served to focus wave energy onto the shore, thus increasing the size and power of the wave. All down Sri Lanka's east coast, this same alternating pattern of submarine ridges and canyons led to the same alternating pattern of death and destruction: great damage here, less damage there, some villages washed away, others hardly affected. As Bruce Parker observes, it is hard to imagine how survivors from these neighbouring villages could ever reconcile themselves to the fact that their fortunes were ultimately determined by the shape of their local sea floor.[14]

Submarine earthquakes are not the only tsunamigenic hazard. In fact the biggest wave ever recorded was caused by a rockfall in an Alaskan fjord. On 9 July 1958 a magnitude 8.3 earthquake along Alaska's Fairweather Fault loosened some 30 million cubic metres (40 million cubic yards) of rock and ice above the head of Lituya Bay, a mountain-fringed inlet on the Gulf of Alaska. The rocks fell directly into the water from a height of nearly a kilometre, precipitating a monstrous wave that reared an astonishing 524 m (1,720 ft), higher than the Empire State Building. The wave lost height as it surged towards the mouth of the 11-km

(7-mile) fjord, scouring millions of trees from the banks as it crashed through the bay. One eyewitness, interviewed the following day by Don J. Miller of the United States Geological Survey, described the sound of the wave as 'an explosion', and recalled the mayday message that he had sent as the wave bore down on his fishing boat, the *Edrie*: 'Mayday! Mayday – *Edrie* in Lituya Bay – all hell broke loose – I think we've had it – goodbye.'[15] Of the three boats in the bay at the time, two had

This aerial photograph, taken by the U.S. Geological Service a few weeks after the Lituya Bay tsunami, shows where the shoreline was stripped of trees by a giant wave.

Close-up of the wave damage in Lituya Bay caused by the 1958 tsunami.

remarkably lucky escapes, while the third, a 55-ft trawler (the *Sunmore*), was sunk with the loss of all hands.

In the years following the megatsunami Lituya Bay's lost conifers were replaced with alders, so every autumn, as the leaves are shed, a line of brown emerges from the surrounding green to reveal the fearsome dimensions of history's biggest wave. But will such a monster ever be seen again? In 1999, a team of geophysicists from University College London made headlines around the world when they predicted the imminent collapse of Cumbre Vieja, an ancient volcano on one of the Canary Islands: its fall, they claimed, would precipitate an Atlantic-wide megatsunami, a 100-m (330-ft) wave that would cause widespread coastal damage throughout Africa and Europe, before going on to destroy much of the populous eastern seaboard of the United States and Brazil. Though the claims remain contentious – oceanographers have since pointed out that Canary Island landslides have tended to be gradual rather than catastrophic – the story did much to raise awareness of the reality of tsunami hazards in oceans other than the Pacific.

Rockfalls have also been responsible for generating sizeable freshwater tsunamis. Research in Switzerland has revealed the surprising extent of a tsunami on Lake Geneva in the sixth century AD, precipitated by a huge Alpine rockfall in the mountains near Geneva. The 8-m (25-ft) wave destroyed the Geneva bridge and swept away villages on the lake shore, causing numerous casualties; a reminder that tsunamis are not restricted to the waters of oceans and fjords.[16]

As was seen in the previous chapter, the strength of a tsunami does not always correspond to the strength of its originating earthquake. A particular kind of slow but powerful undersea rupture, known as a tsunami earthquake, can trigger tsunamis of a far greater magnitude than the earthquake itself, as happened in Japan in 1896. These long-period quakes remain particularly dangerous, as they are usually too weak to generate automated tsunami warnings, and even if the coastal inhabitants are aware of the minor tremor, they will be unlikely to expect a destructive tsunami to follow.

The 1896 Sanriku disaster was a case in point, and its shocking severity led to a new era of seismology in Japan, where the horizontal seismograph had just been invented by a team led by a visiting British geologist, John Milne. In 1899 seismologist Akitsune Imamura proposed that the tsunami had been triggered by movements of the earth's crust under the sea, anticipating the later theory of plate tectonics. As had John Michell in the previous century, Imamura studied historical earthquake patterns, and from them he extrapolated that a major earthquake was due to hit the Tokyo area within 50 years, resulting in enormous loss of life. In a paper published in 1905 he urged the Japanese government to invest in a range of antiseismic measures across the region but, predictably, little notice was taken. His warning was borne out in 1923, when the Great Kantō earthquake and tsunami devastated Tokyo, killing more than 140,000 people. Since then, Japan has led the world in antiseismic building design although, as was seen in March 2011, the best-made structures on the planet can offer little resistance to the unimaginable power of the sea.

The word tsunami

It was the 1896 disaster that introduced the Anglophone world to the Japanese word 'tsunami'. In Japan itself the term dates back to at least the seventeenth century, with its first written appearance in a journal kept by an anonymous court employee. The entry makes reference to an earthquake and tsunami along the Sanriku coast in December 1611: 'It transpired that Masamune's land by the sea was hit by towering waves and all properties were lost. 5000 died of drowning. People call it a tsunami.'[17] The word was evidently already in use among coastal populations, and seems likely to have been coined by fishermen who would have called the phenomenon 'tsunami' ('harbour wave') because such sea-waves only become visible as they enter coastal waters.

Earlier Japanese accounts of tsunamis used a range of different terms. One seventh-century court document described the 684 Hakuho Nankai tsunami as an *oshio* ('large tide'), while other pre-'tsunami' usages include *onami* ('large wave'); *shikai namisu* ('waves rise in all directions'); *takanami* ('high wave'); *takashio* ('high tide'); and *kaisho* ('roaring sea'), variants that make it difficult to know whether a particular historical inundation was truly the result of a tsunami.[18] The term 'supertidal wave', meanwhile, had been introduced to science in the wake of the Krakatoa eruption, in one of the Royal Society's many published reports on the disaster, with the withdrawal of the sea referred to as a 'negative supertidal wave'.[19]

The word 'tsunami' entered the English language through the work of two travellers who published reports of the Meiji Sanriku tsunami. The first was by the American travel writer Eliza Ruhamah Scidmore, who was a frequent visitor to Japan in the 1880s and '90s (her brother was a diplomat based in Yokohama), and was later instrumental in the planting of Japanese cherry trees along the banks of the Potomac River in Washington. Her account of the Sanriku event was published in the *National Geographic* magazine for September 1896, in which she reported that

> on the evening of 15 June 1896 the north-east coast of Hondo, the main island of Japan, was struck by a great earthquake wave (tsunami), which was more destructive of life and property than any earthquake convulsion of this century in that empire.[20]

The entire Sanriku coastline, she wrote, had been 'laid waste by a great wave moving from the east and south, that varied in recorded height from 10 to 50 feet'.

Three months later, a more detailed account of the disaster was published by the Greek–Irish ethnomythologist Lafcadio Hearn, a wanderer who had settled in Japan and taken the name Koizumi Yakumo. Hearn witnessed the aftermath of the tsunami, which he described in an article for the *Atlantic Monthly* in December 1896. 'From immemorial time', he wrote, 'the shores of Japan have been swept, at irregular intervals of centuries, by enormous tidal waves, – tidal waves caused by earthquakes or by submarine volcanic action. These awful sudden risings of the sea are called by the Japanese *tsunami*.' Hearn's mesmerizing description of the wave was based on details derived from eyewitness testimony:

> Through the twilight eastward all looked, and saw at the edge of the dusky horizon a long, lean, dim line like the shadowing of a coast where no coast ever was, – a line that thickened as they gazed, that broadened as a coast-line broadens to the eyes of one approaching it, yet incomparably more quickly. For that long darkness was the returning sea, towering like a cliff, and coursing more swiftly than the kite flies. '*Tsunami!*' shrieked the people; and then all shrieks and all sounds and all power to hear sounds were annihilated by a nameless shock heavier than any thunder, as the colossal swell smote the shore with a weight that sent a shudder through the hills.[21]

Following these initial appearances, the word 'tsunami' began to appear elsewhere in English-language contexts. An article on seismology in Japan that was published in the journal *Nature* in 1905, for example, noted how 'out of forty-seven destructive earthquakes which originated beneath the Pacific, twenty-three were accompanied by *tsunami* or sea waves'.[22] But the usage was not widespread, and it was only after the 1946 April Fool's Day tsunami that the term began to be widely taken up, particularly among seismologists such as Francis Shepard, in preference to the older and more familiar 'tidal wave', as well as to the Royal Society's proposed 'supertidal wave'.

By the early twenty-first century, following the globally reported 1998 Papua New Guinea tsunami, the word 'tsunami' had become established enough to begin to take on figurative usage, and was already being applied to an array of metaphorical inundations. A quick search of the online archives of British and American newspapers turns up, among many other examples, 'a tsunami of bad news'; 'a tsunami of disbelieving joy' (in the wake of the England football team's 5–1 win over Germany in September 2001); 'a tsunami of obesity'; 'a tsunami of junk food' (suffered by journalists on the u.s. campaign trail); 'the ageing-population tsunami'; and even, in the headline of an article in the *Lancet*, 'a global tsunami of cardiovascular disease'.[23] And some of my teaching colleagues regularly refer to the tsunami of marking that washes onto their desks every summer. But while 'tsunami' has now been fully naturalized into figurative English, it is interesting to note that 'tidal wave' has all but disappeared as a metaphor.

Historical tsunami research

Seismology is a science of prediction, which means, in practice, that most of its efforts are directed towards reconstructing the past, seeking out historical clues that might help prepare for the future. Palaeoseismology is a particularly fast-developing branch of earth science that looks at historic and prehistoric events for which there are either no written records – as in the case of the Storrega slide megatsunami of around 8,000 years ago, which is thought to have slammed into Scotland and eastern England from a submarine landslide off the coast of Norway – or where the written records lack reliability, as in the case of the eastern Mediterranean earthquake and tsunami that struck the Egyptian city of Alexandria in AD 365.

The latter event was described in great detail by the fourth-century Roman historian Ammianus Marcellinus in Book XXVI of his *Res Gestae* ('Things Done'). Just after daybreak on 21 July, he wrote:

> the sea was driven away, its waves were rolled back, and it
> disappeared, so that the abyss of the depths was uncovered
> and many-shaped varieties of sea-creatures were seen
> stuck in the slime . . . people wandered at will about the
> paltry remains of the waters to collect fish and the like
> in their hands; then the roaring sea as if insulted by its
> repulse rises back in turn, and through the teeming shoals

dashed itself violently on islands and extensive tracts of
the mainland, and flattened innumerable buildings.[24]

Marcellinus's account went on to record thousands of deaths
by drowning, with bodies found along the length of the coast,
and ships left stranded on the roofs of houses 'nearly two miles
from the shore'.

At first glance this is evidently a classic description of a
destructive tsunami, complete with a verifiable date and time;
indeed, Alexandria's 'day of horror', 21 July, went on to be
commemorated annually until well into the sixth century. But
there are two kinds of evidence to which close attention must be
paid: historical and geological, both of which are riddled with
uncertainties. Firstly, it is never clear whether an ancient source
is describing a single event or an amalgamation of several events
over time. It is always possible that Marcellinus was in fact
writing a portmanteau account of an extensive period of seismicity, condensed into a single potent episode. Secondly,
geologists had long been unable to discover corroborating
ground evidence, such as rupture sites or coral deposits, that
might confirm the date and description of the earthquake and
tsunami. In 2008, however, a team of researchers led by British
geologist Beth Shaw finally located a fault along the southern
edge of Crete that had uplifted the western part of the island by
up to 10 m (32 ft), leaving marks 'that resembled a bath-ring
around the coastline'.[25] The same mechanism that lifted Crete
would also have lifted the sea floor, generating a sizeable tsunami
across much of the eastern Mediterranean. Carbon-dating of
the exposed Cretan corals, meanwhile, confirmed that the event
had happened around the time that Marcellinus and other
historians had claimed. Thus the written records were borne out
by the geological evidence.

The same research also revealed that the fault continues to
accumulate strain energy today, with computer models and field
evidence suggesting that a similar-sized earthquake and tsunami
occurs in the region roughly every 800 years. Given that the last
in the area occurred in 1303, there is clearly a need to educate

the growing coastal populations of the Mediterranean in what to expect and how to react when the next major earthquake occurs. The challenge, as always, is to convince regional policy-makers to listen to what the scientists are saying.

Tsunamis in the past were often misidentified as floods or storm surges, especially if an originating earthquake or landslide occurred at too great a distance for the events to appear causally connected. A particularly striking example of this was the Bristol Channel flood of 30 January 1607, which had long been assumed to be a catastrophic storm surge, in spite of the fact that several eyewitness accounts described the weather as having been sunny and fine. The advancing wave, which one observer described as 'huge and mighty hilles of water tumbling over one another in such sort as if the greatest mountains in the world had overwhelmed the low villages or marshy grounds', was, significantly, preceded by the waters of the Bristol Channel appearing to be 'driven back' – the characteristic herald of an approaching tsunami.[26]

Field research undertaken by geologists Simon Haslett and Edward Bryant lends support to the idea that this fearsome inundation, which drowned more than 2,000 people 'with the greatest violence', was indeed a tsunami, caused in all likelihood by an undersea rupture along a known fault off the coast of Ireland, where records confirm a tremor had been felt that morning. Evidence on the ground includes large boulders displaced up the beach by waves of enormous force; a 20-cm (8-in.) layer of sand, shells and stones discovered within an otherwise unvaried deposit of mud along the coast; and rock erosion characteristic of high water velocities present throughout the Severn Estuary. Putting these apparent historic tsunami signals together with the surviving eyewitness accounts that describe the frothing wave advancing under a clear sky, 'with such a smoke as if mountains were all on fire ... as if myriads of thousands of arrows had been shot forth all at one time', it seems hard to conclude that this traumatic event was anything other than a tsunami.[27]

Palaeoseismology has solved many such mysteries in recent years, including the enduring enigma of the so-called 'orphan

The Bristol Channel flood (tsunami?) of 1607, from a contemporary pamphlet entitled *Lamentable newes out of Monmouthshire in Wales* (1607). Around 2,000 people were drowned by the waves, which travelled 'with a swiftness so incredible, as that no gray-hounde could have escaped by running before them'.

tsunami' of January 1700, which struck the east coast of Japan with no preceding earthquake experienced anywhere in the western Pacific. The relationship between earthquakes and tsunamis was already well understood in Japan, but the event remained a mystery for nearly 300 years, and was widely thought to have been an unusually catastrophic high tide rather than a true tsunami. In the 1980s, however, geologists working in the Pacific Northwest began to piece together evidence from coastal Oregon and California, which run alongside the highly active Cascadia fault some 80 km (50 miles) offshore. Traces of numerous historic earthquakes and tsunamis had been found along much of this Cascadia coastline, with one particularly catastrophic event datable to the turn of the eighteenth century. The most compelling evidence came from tree-ring studies showing that cedars and spruce trees killed by the sudden submergence of coastal forests had outermost growth rings formed in 1699, the last growing season before the 'orphan tsunami' struck. This established the time frame of the earthquake and associated subsidence to the growing season between August 1699 and May 1700.

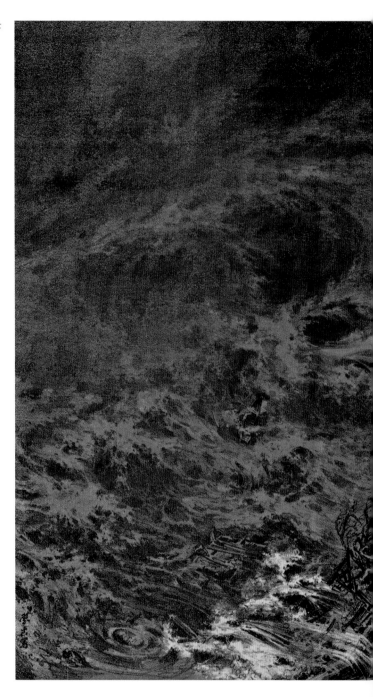

A 19th-century print showing a tsunami battering the coast of Iwate Prefecture, Japan.

Historical records from Japan, meanwhile, dated the arrival of the tsunami more precisely to the early hours of 26 January 1700. Putting the various pieces of historical and geological evidence together made the case for a magnitude 8.7 to 9.2 megathrust earthquake that ruptured some 1,000 km (620 miles) of the Cascadia subduction zone at 9 o'clock in the evening, precipitating a powerful trans-Pacific tsunami that crashed without warning into northeast Japan some eight hours later. The orphan's parent earthquake had been found.[28]

Again, this solving of a riddle from the past also served to shine a light into the future. Tsunamigenic earthquakes occur along the Cascadia subduction zone roughly every 500 years; with warning systems now in place across the Pacific, Japan is unlikely to be taken by surprise by a long-range tsunami from the next Cascadia earthquake, whenever it may come, but what of North America itself? How prepared are the big cities along the Cascadia coast (Portland, Seattle, Vancouver) for a repeat of what we know from Native American accounts to have been a sudden and destructive inundation? Not very, according to Jerry Thompson, author of *Cascadia's Fault: The Coming Earthquake and Tsunami that Could Devastate North America* (2011), who notes that geologists estimate a 37 per cent chance of a major (magnitude 8 to 9) megathrust earthquake occurring within the next 50 years. The consequences could easily match what happened in Japan in March 2011, although the fact that Japan leads the world in disaster preparedness, yet still suffered to the extent that it did, does not bode well for the relatively un-prepared Pacific Northwest. On 11 March 2013 – the second anniversary of the Japanese tsunami – the Oregon Resilience Plan was unveiled at a press conference in Portland. Its sobering conclusion, that 'Oregon is far from resilient to the impact of a great Cascadia earthquake today', was underpinned by the disquieting fact that there would be as little as fifteen minutes between an earthquake and the arrival of the subsequent tsunami: enough time to issue a wave warning, perhaps, but not enough time to heed it. Geologist Bryan Atwater has summed up the basics of tsunami survival in a stark three-line to do list, which every coastal resident in the world, especially those who live along subduction zones, should keep at the forefront of their minds:

> If you feel a strong earthquake, run to high ground. If the sea recedes strangely, run to high ground. If a tsunami ensues, stay on high ground; its first wave probably won't be the last – or the highest.[29]

Tsunami bomb

Not all tsunami science has been devoted to preserving life. In 1944, during the closing act of the Second World War, a top secret experiment was conducted by the New Zealand military, code-named 'Project Seal'. Originally intended as a backup to the planned atomic bomb attacks, the experiment sought to discover whether an offshore explosion could create a synthetic tsunami big enough to inundate the Japanese coast. The idea was first mooted by an American naval officer who had observed military blasting operations around Pacific island reefs. Some of the explosions, he noticed, produced unexpectedly large waves: could this observation not be put to military use? His idea was passed to the New Zealand War Cabinet, who agreed to sponsor preliminary tests off the islands of New Caledonia.

According to the Maori filmmaker and historian Ray Waru, who unearthed the hitherto classified story in New Zealand's national archives in 2011, the initial experiments proved encouraging enough for the allied forces to take the proposed 'tsunami bomb' seriously. 'In my opinion', wrote Admiral William F. Halsey, the u.s. Pacific commander, 'inundation in amphibious warfare has definite and far-reaching possibilities as an offensive weapon.'[30] With Halsey's encouragement, the New Zealand War Cabinet authorized a further series of tests to be conducted off the Whangaparaoa Peninsula, near Auckland. In all, nearly 4,000 undersea explosions were carried out, using a range of different charges, but none succeeded in generating a sizeable wave. It was soon realized that a series of enormous synchronized explosions would be needed to do the job, 'a line or array of massive charges totalling as much as two million kilograms', to be detonated about 8 km (5 miles) from shore. This, it was calculated, had the potential to produce a destructive wave around 10–12 m (32–40 ft) high, big enough to knock out some of Japan's coastal defences. But the idea was never put into practice. By early 1945 the atom bomb was all but ready, and the Allies were prevailing in the Pacific. Project Seal was closed down, leaving the military potential of a tsunami bomb unrealized.

There are suspicions, however, that Project Seal may in fact have been an elaborate hoax, an intelligence ruse designed to distract Soviet attention from the imminent bombing of Japan. Whatever the truth (and much of the Project's documentation remains restricted) it seemed appropriate to end this discussion of tsunami science with something that may well turn out to be tsunami science fiction: an apt introduction to the following chapter, which will look at the representation of tsunamis in global myth and legend.

3 'The Hungry Wave': Tsunamis in Myth and Legend

In a Yurok myth recorded a century ago, Thunder wants people to have enough to eat. He thinks they will if prairies can be made into ocean. He asks Earthquake for help. Earthquake runs about, land sinks, and prairies become ocean teeming with salmon, seals and whales.[1]

In early January 2005, a week after the Boxing Day tsunami, it was rumoured that the body of a dead mermaid had been found on a beach in southern India. Though clearly untrue, it was, nonetheless, a poignant story; not so much a hoax as a mythic response to the unsettling fact that bodies recovered from the aftermath of tsunamis are usually in no condition to be identified. Battered, bloated and often unrecognizable as human, they might as well be thought of as mermaids wrenched from their ancestral element, in a piece of magical thinking designed to mask the pain of reality.

Such stories are a characteristic feature of seismically active regions. Wherever there are tsunamis there are tsunami legends, and this chapter will look at a range of examples from around the world, considering them in their cultural, historical and geological contexts.

Cautionary tales

One of the most remarkable survival stories to emerge in the wake of the Indian Ocean tsunami concerned the isolated Moken people – semi-nomadic, sea-based Austronesians who fish the waters of the Andaman Islands off the west coasts of Burma and Thailand. The Andamans were hit hard by the tsunami within half an hour of the megathrust earthquake, but though most of their homes and many of their boats were destroyed by the giant waves, only one among the 3,000 or so

Moken 'Sea Gypsies' was reported to have died that day. At the first sign of the sea's withdrawal they had either headed inland to safety, or paddled out to deep water in their traditional *kabang*s. It was their knowledge of the sea that saved them from the tsunami: knowledge handed down over many generations in the form of cautionary fireside tales such as the legend of the hungry wave.

The legend tells how, from time to time, the La-boon ('the hungry wave') is invoked by angry ancestral spirits, whose fury is so great it shakes the earth. Before the wave arrives to feed on those who have failed to heed the warning signs, the sea retreats from the shore: so fierce is the La-boon's hunger that other waves flee in terror. It is then that the famished waters arrive, flooding the earth and consuming everything in their path.

The story may be a familiar flood myth of elemental destruction and renewal – the world reborn after a vengeful inundation – but it is also an accurate description of the sea's behaviour, in which the ocean apparently empties itself before the arrival of a giant wave. As the American schoolteacher Marsue McGinnis had observed in 1946, 'it was as if the ocean had taken a deep, belly-filling mouthful of its own water'.[2] Thus the Moken's memorable folk tale, in which disaster is foreshadowed by a readable omen, served as an effective tsunami warning system.

A similar story tradition saved most of the inhabitants of Simeulue, a mountainous island 150 km (93 miles) off the west coast of Sumatra, and only a few kilometres from the earthquake's epicentre. Within minutes of the powerful quake the island's northwest shore was buffeted by tsunami waves between 2 and 10 m (6 and 30 ft) high, but of a 75,000-strong population only seven people were reported to have been killed. The remarkably low death toll on what was the nearest inhabited coastline to the earthquake's strike zone can also be attributed to the cautionary tales that are part of the island's folklore. In Simeulue's case, the stories can be traced back to January 1907, when a powerful local tsunami (originating from the same stretch of fault line as the 2004 event) killed

A traditional *kabang* in which the semi-nomadic Moken people of the Andaman Islands spend six months of the year at sea. Their intimate knowledge of ocean behaviour, preserved in stories such as the 'hungry wave', saved their lives during the 2004 Indian Ocean tsunami.

around half of the island's population. Many were drowned when they ran to the beach after seeing the water recede. As they busily gathered stranded fish, the killer wave powered towards them.

Those who survived made sure to teach their children and grandchildren about the dangers of the *semong* – the local word for tsunami – in the form of stories and songs. One describes how the ocean behaves like a giant bathtub, in which sudden movements generate powerful waves that can travel at dangerous speeds, while another compares the ocean to a see-saw that first tilts down, and then suddenly rushes up: an instructive image born from bitter knowledge of the sea. It was because of this strong oral tradition that the Simeulueans knew what to do when the Boxing Day earthquake struck. After the ground stopped shaking, and the sea began to withdraw, cries of '*semong!*' rang out, at which the majority of islanders fled to the nearby hill slopes before the first of the wave train arrived.[3] Without such awareness rooted deep in their culture, the islanders would have died in their thousands once more.

Stories, however, offer no protection against earthquakes, and a strong aftershock in March 2005 levelled most large structures and claimed hundreds of lives across the islands off northwest Sumatra, including on Simeulue.

Tsunami folklore

In Indonesia's Sunda Strait, where Krakatoa exploded so dramatically in 1883, local fishermen had long complained of being woken at night by the rumbling echoes of the sulphur-breathing mountain spirit Orang Alijeh, or by the sighs of the sea ghost Antoe Laoet, responsible, it was said, for summoning violent waves whenever he was roused to anger.[4] Visiting Dutch surveyors had declared the volcano to be extinct, but did any of them stop to wonder why Krakatoa's earlier inhabitants had abandoned their ghost-ridden island, while its neighbour Sebesi still supported a sizeable population? Given their policy of enlarging the settlements around the strategically important Sunda Strait, perhaps the colonial administration should have paid more attention to those elemental tales of fire and water.

In Hawaii, where volcanoes, earthquakes and tsunamis are equally familiar, oral and written culture is replete with observations derived from centuries of exposure to the sea. The Hawaiian word for tsunami, for example – *kai e'e* – refers specifically to the mountainous appearance of the incoming wave, and

A tsunami breaking on the Hawaiian shore, in a 19th-century woodcut. Hawaii has been battered by many tsunamis over the centuries, as its language and folklore attest.

is distinct from both *kai hohonu*, the Hawaiian term for a high tide, and *kai ea*, an unusually fast-rising sea. The word that describes the characteristic withdrawal of the water that occurs in advance of the wave, meanwhile, is *kai mimiki*. There is even a term, *kai ku piki'o*, that describes the seismic seiche (the agitation of enclosed bodies of water that is often observed after an earthquake).[5] This is a language that bears eloquent witness to life lived at the mercy of the sea.

Hawaiian folk tales bear similar witness to long familiarity with tsunamis. One story tells of an illicit love affair between a Big Island chieftain, King Konikonia, and a mer-woman who lived in the sea off Hilo Bay. The woman was persuaded ashore by the love-struck king, who took her to live in his waterside palace, but the next day she warned him that her angry brothers (in the form of blennie fish) were likely to come and take her home by force. In order to make their way across the beach to the palace, the brother-fish would seek help from an incoming sea. Ten days later, according to the legend, 'the ocean rose and overwhelmed the land from one end to the other' until it reached the door of King Konikonia's palace. From there the mer-woman was seized by a shoal of her avenging brothers, while many of Konikonia's unfortunate subjects were swept out to sea to their deaths. Hilo Bay was abandoned, but some years later survivors of the king's tsunami began to return and resettle the area, vowing never to anger the ocean or its inhabitants again.[6]

A similar story of human folly and repentance is narrated by an ancient flood myth from Sri Lanka: King Kelanitissa of Kelaniya (now a suburb of the island's modern capital, Colombo) once ordered an innocent monk to be boiled alive, for which the gods, angered by such cruelty, made the ocean rush inland and flood the coastal kingdom. The king's advisors claimed that if a princess was sacrificed to the sea, the incoming waves would stop. To the king's dismay, his eldest daughter volunteered herself as a sacrifice: she was placed inside a beautifully decorated boat which bore the words 'Daughter of a King', and set adrift on the turbulent sea. As soon as the vessel was launched the waves died down and the floodwaters receded,

Viharamahadevi, Kirinda, southern Sri Lanka, overlooking
the spot where the princess's boat was reputed to have
made landfall.

although the king and his subjects remained distraught at the loss of the brave princess – who happily survived her post-tsunami sea voyage and married a neighbouring (and more kindly) king, Kavantissa of Ruhuna. The princess took the name Viharamahadevi from the spot where her boat made landfall, near the Lanka Vihara, a revered Buddhist rock temple, where a statue of her now gazes serenely out to sea. She remains exalted as one of the great legendary heroines of Sri Lanka.

A cautionary tale from Japan, meanwhile, told the story of a samurai who survived the 1611 Sanriku tsunami by faithfully serving his master, Masamune. The master had wanted fish, so dispatched two samurai to round up some fishermen, but the men were frightened by the sea's strange appearance and refused to leave the shore. One of the samurai insisted on obeying his master's order, and compelled the fishermen to launch their boat:

> soon it meets the tsunami, which drives it inland into the crown of a pine tree. The waves also sweep away entire villages along the shore. Later, after the water recedes, the men clamber down from the tree. Scanning the shore, they realize that they too would have been swept away had they not gone fishing for Masamune.[7]

The samurai was richly rewarded for his devotion – thus the moral of the story is: if you follow orders, you may escape disaster and receive the blessings of your master.

What is so impressive about such legends, at least from a modern Western perspective, is their haunting combination of the magical and the real, typified by the way they offer animistic or divine explanations for otherwise precisely observed phenomena. They act as a narrative bridge between the actual and the imagined, though in the case of one tsunami legend from a neighbouring stretch of coastline, the imagined in the story turns out to have been more actual than it first appeared. The story in question, from Tamil Nadu in southern India, told how Indra, the Hindu god of rain and thunderstorms, grew jealous of the beauty of the seven pagodas of the famous temple complex at Mahabalipuram, a former seaport on the Bay of Bengal some 60 km (37 miles) south of the city of Chennai. The god sent a great wave to destroy the temples, leaving only a single building intact – the intricately decorated Shore Temple, now a UNESCO World Heritage Site. But stories of the lost pagodas persisted in Tamil legend, some of which were recorded and published in the late eighteenth century by John Goldingham of the Madras Observatory. One local Brahmin, interviewed by Goldingham in 1798, recalled that 'his grandfather had frequently mentioned having seen the gilt tops of five pagodas in the surf, no longer visible'.[8]

Little credence was given to such stories until the morning of Boxing Day 2004, when the deep withdrawal of the water in advance of the first wave exposed what appeared to be ancient granite remains sunk into the seabed around half a kilometre from shore, though no photographs of them are known to have been taken. A few minutes later the incoming wave resubmerged them, before striking (and, as it turned out, thoroughly cleaning) the surviving Shore Temple as it did so. No major structural damage was sustained to the temple due to the strength of its granite foundations, but the tsunami was about to introduce another twist to the story. Its huge undertow scoured large quantities of sand and soil from the beachfront, exposing an array of previously unknown carved remains, as well as a temple-like structure resembling parts of a Hindu shrine. The nearly 2-m-high (6-ft) carvings have been dated to around the seventh century AD, and their bold designs, including a lion, a horse and an elephant, are typical of the art that thrived during southern India's Pallava era. The tsunami's sudden de-silting of the site has presented archaeologists with some puzzles to solve: are these structures the remains of a larger temple complex – perhaps even the legendary Seven Pagodas – or are they part of the long-vanished seaport that flourished in the area more than 1,200 years ago? And was the site really destroyed by an earlier tsunami, as the local legend suggests? Layers of sea shells and oceanic debris beyond the tideline point to possible historic tsunami

The Shore Temple, Mahabalipuram, southern India, which, according to local tradition is the sole survivor of the legendary Seven Pagodas destroyed by an ancient tsunami.

inundation, a picture corroborated by similar findings further up the Coromandel Coast. The mystery of Mahabalipuram deepens.[9]

Another mystery, from another part of the world this time, offers a useful reminder that tsunamis are not exclusively seismic in origin. A collection of Aboriginal and Maori legends appears to describe a tsunamigenic meteorite strike – the so-called Mahuika comet, named after the Maori god of fire – that occurred somewhere in the southwest Pacific towards the end of the fifteenth century. Stories from New Zealand's South Island tell of 'the falling of the skies, raging winds, and mysterious and massive firestorms from space', followed by a catastrophic deluge 'that flooded the Aparima Plains west of Invercargill'.[10] Datable tsunami deposits along much of the coastline, as well the incidence of inland place names featuring the Maori word *tai*, or 'wave' (e.g. Tainui; Tairoa; Paretai) lend credence to the theory that the skyfall myths invoke a powerful cosmogenic tsunami that was witnessed by the islanders several centuries ago: the same event, perhaps, that southeast Australian Aborigines also

describe in a host of similar legends that tell how 'a spear from the sky fell into the sea followed by a great flood that changed the coastline'. According to geologist Edward Bryant, who has conducted many years of fieldwork in the region, South Australia's coastline is marked by clear tsunami signatures, including wave-thrown boulders that feature in numerous local legends, such as the story of Ngurunderi, an important ancestral figure for some of South Australia's coastal tribes. The story tells how Ngurunderi's two wives ran away from him; he set off in angry pursuit, eventually catching up with them at the tip of the Fleurieu Peninsula, just as they were attempting to wade from Kangaroo Island to the mainland:

> To punish them he ordered the waters to rise up as a tidal wave and drown them. The waters came in with a terrific rush and roar, carrying the women toward the mainland. They tried to swim against the wave, but were drowned; their bodies were turned to stone and are seen as two rocks off the coast of Cape Jervis, called the *Pages* or the *Two Sisters*.[11]

Excavation pits dug by the Archaeological Survey of India, just south of the Shore Temple, Mahabalipuram, where a submerged structure was revealed by the 2004 tsunami – perhaps part of the ancient seaport that once flourished on the site?

Another Aboriginal inundation story tells how an ancestor, Tibrogargan, was alarmed to see a sudden great rising of the ocean, and fled inland to safety with his family. A range of peaks in the Glasshouse Mountains, some 20 km (12 miles) from the shore, are said to represent the family still gazing seawards at the threat.

As all these stories attest, tsunamis and tsunami memories are prolific generators of myth and legend: a natural response, perhaps, to their primal ferocity, to the trauma of witnessing the sea rise up against the land as though the very elements were at war. Indeed, many tsunami stories stress the warlike nature of this elemental collision, such as the Native American legend of Thunderbird and Whale, which ethnologists date to around 1700, the date of a cataclysmic earthquake and tsunami generated by an undersea rupture along the Pacific northwest coast. That same 1700 Cascadia earthquake was the source of the notorious trans-Pacific tsunami known to the Japanese as 'the orphan tsunami', discussed in the previous chapter.

Rock carvings, exposed by the tsunami at Mahabalipuram, showing a horse and an elephant, typical of the art of the Pallava era (4th to 9th centuries AD).

Detail of the Thunderbird and Killer Whale Totem Pole by Harold Alfred.

Thunderbird and Whale stories tell of two elemental beings of supernatural size and power, locked in a battle to the death. Variations of the myth have featured in the oral traditions of Pacific Northwest peoples along the entire Cascadia coast, from British Columbia to northern California. In one version, all earthly creation is said to rest on the back of a giant killer whale, while Thunderbird causes thunder overhead by ruffling his feathers and pouring down rain from a large lake carried on his back. The vengeful (though, in most of these stories, benevolent) Thunderbird sometimes drives his talons into Whale's back, causing Whale to writhe and dive, dragging the struggling Thunderbird down to the bottom of the sea. Their thrashing leads to terrible earthquakes and floods on land.

At least one version of the story, from the Olympic Peninsula in Washington State, draws an explicit connection between this epic, elemental battle of the giants and a powerful tsunamigenic earthquake:

Spindle whorl, carved in relief with a Thunderbird and Whale design. Coast Salish people, Pacific Northwest.

following the killing of this destroyer . . . there was a great storm and hail and flashes of lightning in the darkened, blackened sky and great and crashing 'thunder-noise' everywhere. There were also a shaking, jumping up and trembling of the earth beneath, and a rolling up of the great waters.[12]

Other coastal stories narrate earthquakes and marine flooding in purely symbolic terms, though there are details that hint at pre-tsunami withdrawal: 'when it lifted its tail and struck the water, the bay became dry. That was the way it had drowned the other people', while another account develops what reads like a mythical explanation for seismic aftershocks, noting that Whale had a son, Subbus, whom Thunderbird took several more days to locate and kill. Their earth-shaking struggles persisted, but eventually Subbus, too, was defeated.

Can these seemingly fanciful myths and legends really be connected to historical earthquakes and tsunamis? A wealth of oral testimony suggests they can: Chief Louis Nookmis, an 84-year-old tribal leader from Vancouver Island, who was interviewed in 1964 in the wake of the Alaska tsunami, recalled 'stories from my grandfather's father (born *c.* 1800) about events that took place four generations before *his* time', and he spoke in vivid terms of the destruction of his ancestral village on the island's west coast by what must have been the Cascadia tsunami of nearly 300 years before:

> They had practically no way or time to try to save themselves. I think it was at nighttime that the land shook . . . I think a big wave smashed into the beach. The Pachena Bay people were lost. But they who lived at Ma:lts'a:s, 'House-Up-Against-Hill' the wave did not reach because they were on high ground . . . Because of that they came out alive. They did not drift out to sea with the others.[13]

Such precisely observed memory narratives, handed down the generations, do much to shed light on the concomitant myths and legends. Stories told to the social historian Beverly Ward in the 1930s by her Northwest Indian grandmother Susan Ned (born in Oregon in 1842), for example, tell of 'a big flood shortly before the white man's time, a huge tidal wave' that struck the Oregon coast:

> The ocean rose up and huge waves swept and surged across the land. Trees were uprooted and villages were swept away. Indians said they tied their canoes to the top of the trees, and some canoes were torn loose and swept away . . . After the tidal wave, the Indians told of tree tops filled with limbs and trash and of finding strange canoes in the woods. The Indians said the big flood and tidal wave tore up the land and changed the rivers. Nobody knows how many Indians died.[14]

Though it can't be known with certainty that it was the 1700 Cascadia tsunami and not some other past event that made its way into these myths and stories, what they add up to is a powerful shared memory of tsunami inundation that has done much to shape the storytelling cultures of the Pacific Northwest coast.

Local knowledge

It is only in recent years that scientists have begun to take such storytelling seriously. Where once they might have dismissed indigenous nature myths as the products of superstition, palaeo-seismologists now tend to seek them out, scouring them for historical pointers to otherwise undocumented events. As Edward Bryant points out, it can be hard to distinguish between fact and fiction, 'between echoes of the real past and dreams', but even the most vague and animistic account of a natural catastrophe can feature precise, even lifesaving, observations.[15] In Papua New Guinea, for example, there is a widely held view that there are no such things as natural disasters, only acts of God or of sorcery by rival tribes. In some places on the island, according to Simon Day, a British geophysicist who has researched historic Papuan tsunamis, 'an earthquake is considered to be a by-product of the spell that a sorcerer uses to cause a tsunami – the two events are related'.[16] Of course such explanations lack scientific credibility but, as has been seen, they can still constitute an effective form of tsunami education. Eyewitness accounts of the 1930 Ninigo Islands tsunami in Papua New Guinea confirm that people in coastal villages recognized the sea's withdrawal as a warning sign, and evacuated inland before the dangerous tsunami struck. They had in mind their grandparents' accounts of the devastating Ritter Island tsunami of 1888, which had been preceded by a sharp withdrawal of the sea and accompanied by a loud roaring sound. When geologist Hugh Davies interviewed survivors of the 1998 tsunami, he was told a number of indigenous tales, such as 'the story of the smoking crab holes of Sissano', in which a great wave comes and buries a village deep beneath the sand. But, he noted, the stories did not include any details of tsunami warning signs, and neither were they widely known: 'only some of the older men who had lived most of their lives in the village, rather than in paid employment elsewhere, recalled that their fathers had told them about tsunamis'.[17]

Both Davies and Day credit traditional tales for the historic low mortality rate among many Pacific island groups, but one of the clearest lessons learned in 1998 was that rapid social and economic change is transforming coastal populations across much of the developing world, and that hard-won tsunami lore is being lost as a result. The challenge today is to develop tsunami-smart knowledge and behaviour among newer immigrant and transient populations that are settling in tsunami-prone areas. Listening to long-established indigenous traditions may well be the best place to begin.

In Japan, for example, the story of how Hamaguchi Gohei saved hundreds of his neighbours from an incoming wave by setting fire to his rice-stacks is still taught in schools across the country. The story, which originated during the 1854 Ansei-Nankai earthquake and tsunami, was made famous outside Japan by Lafcadio Hearn, whom we met earlier in connection with the transmission of the word 'tsunami' into Anglophone usage. In an article entitled 'A Living God', later collected in his luminous book *Gleanings from Buddha-Fields* (1897), Hearn described how one evening in December 1854, Hamaguchi Gohei, the headman of Hirokawa, a village some 70 km (43 miles) south of Osaka, was looking out to sea from his house on the hill, when he felt what he knew to be an earthquake:

> It was not strong enough to frighten anybody; but Hamaguchi, who had felt hundreds of shocks in his time, thought it was queer, – a long, slow, spongy motion. Probably it was but the after-tremor of some immense seismic action very far away. The house crackled and rocked gently several times; then all became still again.[18]

But a little while later he observed the sea begin to darken and change direction: 'it seemed to be moving against the wind. *It was running away from the land . . .*' Soon the rest of the villagers had also noticed the sea's disappearance, and many of them made their way down to the unfamiliar terrain of ribbed sand and wide expanses of weed-hung rock. None of them had seen such a sight before, but Hamaguchi began to remember 'things told him in his childhood by his father's father', and suddenly he realized what the sea was about to do.

The headman knew it would take too long for a warning message to be sent to the village or to arrange for the priest in the nearest temple to toll the alarm bell, so he decided on a more radical course of action. Gohei lit a torch, went out into the fields above the village and proceeded to set fire to the enormous rice stacks that were ready to be transported to market. As soon as his neighbours on the beach saw the flames they rushed inland and up the hill, but to their amazement Hamaguchi forbad anyone to go near the flames until every villager had been assembled. His motives only became clear when he pointed out to sea, where a long, dark line could now be seen approaching the shore, towering like a cliff on the far horizon, 'and coursing more swiftly than the kite flies':

> '*Tsunami!*' shrieked the people; and then all shrieks and all sounds and all power to hear sounds were annihilated by a nameless shock heavier than any thunder, as the colossal swell smote the shore with a weight that sent a shudder through the hills, and with a foam-burst like a blaze of sheet-lightning. Then for an instant nothing was visible but a storm of spray rushing up the slope like a cloud; and the people scattered back in panic from the mere menace of it. When they looked again, they saw a white horror of sea raving over the place of their homes. It drew back roaring, and tearing out the bowels of the land as it went. Twice, thrice, five times the sea struck and ebbed, but each time with lesser surges: then it returned to its ancient bed and stayed, — still raging, as after a typhoon.[19]

The tsunami was the result of an earthquake (estimated magnitude 8.4) along the offshore Nankai Trough, a highly active subduction zone with a long history of generating megathrust earthquakes and tsunamis. Sediment analysis suggests that the height of the main wave was between 5 and 7.5 m (16 and 25 ft), enough to overwhelm a low-lying coastal settlement like Hirogawa, which was almost entirely obliterated, along with most of its surrounding rice terraces. But only a handful of lives were lost in the village, due to Hamaguchi's heroic actions. Many other villages along the coast had not been so fortunate.

In the following months, under Hamaguchi's guidance, a 5-m-high sea wall was built as a protection against future tsunamis. The embankment proved its worth in December 1946, when another damaging tsunamigenic earthquake occurred along the same stretch of the Nankai Trough. This time the waves failed to touch Hirogawa: thus, even many years after his death, Hamaguchi Gohei – whom the villagers had begun to revere as a deity – continued to look after his neighbours. The story of his rice-beacons is taught as a means of passing on lifesaving wisdom to the young.

Animal myths

The Hamaguchi story marked a significant departure from earlier Japanese tsunami legends, which had tended towards the supernatural. Traditional Japanese folk tales had routinely ascribed the cause of earthquakes to the giant catfish Namazu, who lives buried in the mud beneath the main island of Honshu. His movements are restrained by a magical rock held by the thunder god Kashima; but Kashima occasionally loses control of the temperamental fish, whose subterranean writhings (like those of the Native American Whale) lead to earthquakes and tsunamis at the surface. In the aftermath of the Japanese earthquakes of 1854 and 1855, large numbers of people flocked to the Kashima shrine in Ibaraki Prefecture, built, it was believed, over the very rock that subdues the fractious fish. Printed images of Namazu circulated in their thousands, as they still do today following major seismic events, the catfish remaining a ubiquitous feature of Japanese disaster lore. Namazu

Namazu, the giant catfish who lies buried in the mud beneath Japan, is restrained with a magical stone by the thunder god, Kashima. When Kashima's effort fails, the catfish's writhings lead to earthquakes and associated tsunamis.

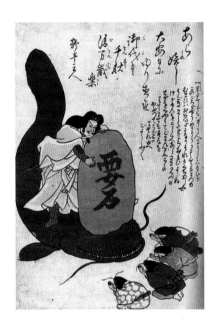

even turns up on modern earthquake warning technology, a testament to how deeply his legend is embedded in the culture.

The Namazu story reflects a once-widespread belief that catfish and carp, with their whiskery barbels, can accurately forecast seismic events. There are many anecdotal accounts from the disastrous Kantō earthquake and tsunami of 1923, when fish in ponds across the city were seen writhing in an agitated fashion in the hours leading up to the quake.[20] Is it possible that bottom-feeding barbelled fish can sense minute changes in electrical currents transmitted underground in advance of a quake; or is their apparent seismic sensitivity connected to an acute sense of taste (catfish are basically tongues with fins) that allows them to detect changes in sediments stirred up by infrasonic waves? Unfortunately, a decade of government-funded research has so far found nothing to support these or any other theories of piscine sensitivity. Indeed, it was reported that not a single research fish displayed unusual behaviour patterns in advance of the March 2011 disaster – at magnitude 9.0, the largest earthquake ever recorded in Japan. Whatever their physiological secrets may be, it is clear that catfish are not about to be inducted into the Japanese seismic warning system.

Land animals have also been credited with similar seismic prescience. The Roman author Claudius Aelianus advised anyone who lived in an earthquake zone to 'watch the animals', and described how in Helike, Greece, in 373 BC, every wild creature quietly left before an earthquake and tsunami destroyed the ill-fated city.[21] The 2004 Boxing Day tsunami gave rise to similar reports of birds and animals seen moving inland before the waves arrived, while dolphins and other marine animals were apparently observed heading further out to sea. Creatures of all sizes, from ants to elephants, featured in these accounts. In Khao Lak, Thailand, a group of eight beach elephants giving morning rides to tourists took off into the jungle just before the wave arrived, thereby saving several lives, an episode on which the Thai writer Tew Bunnag based a short

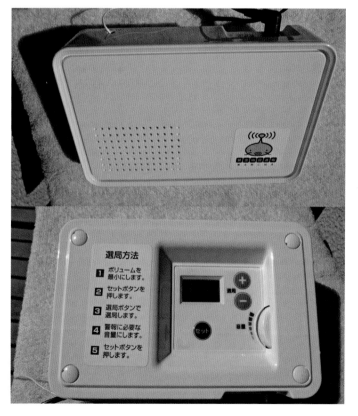

An Earthquake Early Warning radio featuring a Namazu logo (the yellow icon), in which Namazu's shaking tail stands in for the aerial and receiver.

story, 'The Reluctant Mahout' (2005), in which a normally placid female elephant from 'The Dream Elephant Camp' escapes and runs inland on Boxing Day morning.[22] Much the same was said to have happened at the nearby Khao Lak Elephant Trekking Centre, where two elephants trumpeted and broke their chains – 'something they had never done before', according to their handlers – and ran for higher ground, along with four startled but extremely lucky Japanese tourists.[23] At Yala National Park, in southern Sri Lanka, a troop of elephants was also said to have panicked and moved inland some time before the tsunami arrived, much to the consternation of their keepers, while an entire herd of water buffalo in Ranong Province, Thailand, suddenly fled the beach on which they had been quietly grazing: minutes later, the first of the tsunami waves arrived.

Like the Japanese catfish stories, most of these reports are anecdotal, and seismologists and zoologists remain sceptical over claims that animals possess any kind of seismic sixth sense. Animals react to such an array of stimuli – hunger, predation, territorial threat – that it is impossible to know if an individual episode of flight behaviour is attributable to seismic disturbance. In fact the only survey of satellite-collared wild elephants in coastal Sri Lanka revealed no alteration in the herd's activities either before or during the tsunami, its authors concluding that 'the movements of these free-ranging elephants are not consis- tent with flight behavior or other potentially aberrant behaviors attributable to extra-sensory perception or sixth sense, or even with an early response to seismic-borne detection of the earth - quake and tsunami'.[24] But the fact remains that surprisingly few wild animals died in the 2004 tsunami. Among them was a female hippopotamus killed in a river in coastal Kenya. Her orphaned calf, Owen, was taken to a nature reserve outside Mombasa where he was adopted by a 130-year-old Aldabra giant tortoise named Mzee ('old man' in Swahili). The two were reported to be inseparable.[25]

Religious perspectives

The universality of disaster legends suggests that people have always invested natural catastrophes with some form of supernatural significance. Even today, with science all around us, we have a lingering cultural need to attribute meaning to our surroundings, and even the most rational of us can find ourselves wondering about the randomness of random events. In 2004, the Indian Ocean tsunami provoked a remarkable outpouring of religious and quasi-religious commentary, much of which attempted to grapple with theodicy's central question: If God is merciful, why does he allow terrible things to happen to innocent people? Prominent theologians from the world's major faith groups were asked this question by an American journalist, Gary Stern, whose subsequent book, *Can God Intervene? How Religion Explains Natural Disasters* (2007) is a valuable compendium of religious responses to the tsunami, ranging from the fundamentalist: 'He sent it as a punishment. This comes from ignoring His laws'; to the fatalist: 'feel sad for the victims of catastrophe, but know that they will be reborn in future lives'; and the philosophical: 'this has nothing to do with blaming God, if that's what you want to know. It is the result of geography. It is about the earth, the world, the environment, the geography of Indonesia'.[26]

Many of Stern's interviewees argued that the meaning of a natural catastrophe lies not in what caused it, but in how humanity responds. 'There are times', as one Jewish theologian argued, 'when God creates or allows moments like these, natural disasters, in order to test people or allow people to express heroism, kindness, compassion'; while others argued that such events serve partly as a reminder of God's dominance over our lives and partly as a test of the ingenuity with which he has blessed us. 'We should have been able to predict the tsunami', observed the Muslim scholar Sayyid M. Syeed, who credited the survival of Indonesia's tsunami-battered mosques not to divine intervention but to the care with which the faithful had built them. 'It would be a travesty of the divine justice if God destroys

In many of Aceh's coastal towns the mosque was the only building left standing in the wake of the Boxing Day tsunami. Photo taken on 2 January 2005.

the people around the mosque but saves his mosque, or his church', he said.[27] Disasters, in other words, may be evidence of the presence of a divine power, but they are not evidence of divine injustice.

Behind these responses lurks an implicit assumption, shared with many of the indigenous myths, that earthquakes and tsunamis are deliberate acts, as though planetary forces are somehow interested in mankind's day to day affairs. Is it only human vanity that prompts us to imagine that ancient geophysical processes, which began long before the emergence of life on earth, are somehow all about us? And isn't the occasional upheaval the price that mankind is obliged to pay for the otherwise beneficial subterranean forces that generate hot springs and mineral ores, as Immanuel Kant once argued? These questions were first aired more than two centuries ago, in the wake of the Lisbon earthquake, when Britain's royal chaplain, William Warburton, observed that human pride had wildly overestimated the significance of the event:

A drunken man shall work as horrid a desolation with
the kick of his foot against an ant-hill, as subterraneous
air and fermented minerals to a populous city. And if we
take in the universe of things, rather with a philosophic
than a religious eye, where is the difference in point of
real importance between them?[28]

The city of Lisbon, in other words, was of scant significance in
the great scheme of things, and its destruction, though terrible
for the victims, was merely a passing event: a remarkable claim
for an eighteenth-century clergyman to have made, though one
that many Buddhists and Hindus today might recognize,
habituated as they are to taking the long, cosmic view of events.
As the Hindu scholar Vasuhda Narayanan explained, 'the universe lasts for millions and millions of years, so the tsunami is
seen as a little blip; it creates untold suffering – we're not being
insensitive by dismissing the suffering – but *as* an act of destruction, it *is* an act of nature. Why does a tiger eat its prey? We
all suffer the consequences of nature . . .'.[29]

It is fair to say that in the immediate aftermath of the Asian
tsunami most Western commentators made little effort to
engage with the spiritual beliefs of the people most affected by
the event. 'It seemed like a scene from the Bible', as Michael
Dobbs reported in the *Washington Post* on 27 December 2004.
'As the waters rose at an incredible rate, I half expected to catch
sight of Noah's Ark.'[30] But in contrast to biblical theodicy, with
its emphasis on rare, unexpected disasters, South East Asian
spiritualities evolved in the presence of constant geophysical
threat. Given that several hundred minor earthquakes and three
or four major ones (of magnitude 5.0 or above) are recorded in
Indonesia *every day*, those who live along the Java Trench have
long accepted seismic instability as an inescapable part of their
existence – an acceptance reflected in their spiritual outlook, in
which theological hand-wringing has no place. The Buddhist
teaching of *teinen* ('resignation') encourages acceptance rather
than questioning of planetary events, while Hinduism evolved
around the central concepts of impermanence and rebirth.

Though nature is seen as divine, the notion of a vengeful god is not central to either of these faiths. As Vasuhda Narayanan observed, Hinduism teaches that the world exists in a state of constant flux and change, so 'yes, these people do believe in gods, but they do not necessarily attribute all evil deeds to the gods – even if they take part in a healing ritual':

> When the fishermen went back to fish, when they resumed normal activities, the families who had lost many people started off with what's called a propitiating ritual to the ocean ... but it was not to calm down an angry god; it was about restoring harmony, thinking of peace.[31]

Modern myths

As has been seen throughout this chapter, myths and legends, both new and revived, routinely appear in the aftermath of disasters, with storytelling playing a vital role in the process of collective recovery. In the case of especially traumatic disasters, these stories can persist for centuries, reverberating down the years as a form of spoken memory, a bulwark against the dangers of forgetting. As a child in Jamaica in the 1970s (my father was a marine biologist who worked for the island's Ministry of Agriculture and Fisheries), I heard stories of how the voices of the dead could still be heard under the water at Port Royal, the ill-fated harbour town destroyed by an undersea earthquake in 1692. The event had been utterly catastrophic. In the words of one contemporary account, 'in the space of three minutes, Port Royal, the fairest town in all the English plantations, was shaken and shattered to pieces, sunk into and covered, for the greater part, by the sea'.[32] Following the powerful noonday quake, in which the ground 'heaved and swelled like the rolling billows', a sudden subsidence caused much of the town's waterfront area to slide into the sea. Buildings crumbled and capsized, and as the land sank, the sea rose up to invade the ruins, abetted by a powerful tsunami that came not from the open ocean but from inside the bay

The coastal town of Port Royal, Jamaica, was destroyed by an undersea earthquake and tsunami on 7 June 1692. Stories of how church bells can still be heard under the water remain in circulation.

itself. Waves of around 2.5 m (8 ft) struck all along Jamaica's northwest coast, destroying wharfs and shipping, and even pitching the royal frigate, *Swan*, over several houses inland. Two thousand lives were lost in the cataclysm, and two-thirds of Port Royal was sunk and destroyed, including the cemetery where the notorious privateer Henry Morgan – 'the victor of Panama and Maracaibo' – had been buried four years earlier.[33]

Today, much of the old town remains under water, a magnet for divers and treasure hunters as well as a still-potent source of myth and legend. I remember being taken on a school trip to the Port Royal Museum and reading an information panel that claimed that – if you listened quietly – you could hear the bells of St Paul's church, sunk below the waves, still tolling with the tide. More than 300 years after the event, these stories still have the power to impress; the power, indeed, to make a passing animist out of anyone.

And, of course, new myths and legends continue to be created. One striking tale that emerged in the wake of the April Fool's Day tsunami concerned the Kuwahara Store building, one of the only bay-front structures in downtown Hilo to survive the inundation intact. A few days before the tsunami, according to the story, an elderly Hawaiian woman had appeared at the door and asked the owner if he could spare her some food. The shopkeeper, who was well known as a kind-hearted man, sat her down on the bench in front of his shop and presented her with an array of food and drink. After she had eaten, she got up to leave, saying, 'You know, you're a good man. Something real big is going to happen, but you'll be okay.'[34] And she was right: the shop survived, and photographs taken at the time show it standing alone among its flattened neighbours, eerily unscathed by the wave.

The tale, which was told throughout Hawaii in the weeks after the tsunami, is both a miraculous survival story and a variation on the legend of 'the mysterious benefactor', a curious new instance of which emerged in Japan following the March 2011 tsunami. In the weeks leading up to the second anniversary of the disaster, people in the small fishing port of Ishinomaki, Miyagi Prefecture, began to receive gold bars anonymously in the post. The town had been devastated by the tsunami, suffering the highest death toll of any one settlement, and much of the wreckage had still not been cleared by the time the gifts began to arrive. The handful of gold bars so far received – which are thought to be worth at least $250,000 – will contribute to the rebuilding of some of Ishinomaki's worst-hit infrastructure, yet no one has any idea who sent them. No doubt, in the years to come, rumour and hearsay will inflate the number of ingots from

The Kuwahara Store, the only building left standing on Hilo's bayfront after the 1946 April Fool's Day tsunami.

a handful to a vaultful, with the whole town magically touched by the hand of some unseen Midas. But then this is how disaster legends have always arisen, phoenix-like from the ruins, and it's how they will always thrive in the retelling.

4 Tsunamis in Literature, Art and Film

Our vision of a tidal wave was what still today everybody thinks
of: *The Poseidon Adventure*. We thought of the John Hall, Dorothy
Lamour movies where they say, 'Here it comes, here it comes,' and
they climb up a palm tree and all the bad guys are washed out and
the good guys climb down and it's over with.[1]

As hours of mesmerizing YouTube footage confirms, a tsunami
is a powerful visual spectacle, once seen, never forgotten. Indeed,
the sight of that terrifying grey-black tide breaking over the sea
walls of Miyako City in March 2011 has given the world an
abiding new image of the stark ungovernability of nature. For
artists, writers and filmmakers alike, the menacing approach
of a destructive wave has proved an enduringly powerful motif,
replete with all manner of atavistic associations, so it is no surprise
to find that, from Homer to *Hereafter*, seaborne disasters have
played a significant part in the development of our storytelling
culture.

Tsunamis in literature

Homer's hero, Odysseus, faced a litany of natural hazards on his
long journey home from Troy, including the boiling turbulence
of the Straits of Messina which, in this passage from Book XII
of the *Odyssey* (*c.* 800–600 BC), appears strongly reminiscent of
an intermittent volcanic tsunami, complete with the fearsome
withdrawal of the sea:

On one side was Skylla, and on the other side was shining
Charybdis, who made her terrible ebb and flow of the sea's
water. When she vomited it up, like a caldron over a strong
fire, the whole sea would boil up in turbulence . . . but when

in turn again she sucked down the sea's salt water, the
turbulence showed all the inner sea, and the rock around it
groaned terribly, and the ground showed at the sea's bottom,
black with sand; and green fear seized upon my companions.[2]

His companions were right to be afraid: the classical
Mediterranean world had been shaped by such elemental forces,
situated as it was above a major seismic fracture zone, an
earthquake factory running northeastwards under the Aegean,
through Turkey and on to the Black Sea. It was at that far
northern end of the fault line that Jason and his Argonauts
encountered a terrible wave as they approached the narrow
entrance to the Bosphorus. The weather had been fine,
according to Apollonius' *Argonautica* (third century BC), but just
as the ship approached the Symplegades – the legendary
Clashing Rocks – the sea suddenly heaved with 'a terrific roar':

Caverns underneath the crags bellowed as the sea came
surging in. A great wave broke against the cliffs and the
white foam swept high above them. Argo was spun round
as the flood reached her . . . when they were suddenly faced
by a tremendous billow arched like an overhanging rock.
They bent their heads down at the sight, for it seemed
about to fall and overwhelm the ship. But Tiphys just in
time checked her as she plunged forward, and the great
wave slid under her keel.[3]

The ship went on to survive a second 'overhanging wave' that
almost overturned it, until the goddess Athene intervened and
stilled the swaying rocks, thereby calming the waters and allowing the *Argo* safe passage through the strait.

Though the scene recounts a dramatized episode from a legendary quest, the story details are strongly reminiscent of a
tsunamigenic earthquake, a regular occurrence in the eastern
Mediterranean, where clashing rocks and billowing seas spelt
trouble for a seafaring people. Stories of tremendous waves from
tsunamis and storm surges would have circulated throughout

Jason and the Argonauts (c. 1925). Arthur C. Michael's colour lithograph depicts the Argo riding over the 'great wave' at the entrance to the Bosphorus.

the region, with embellishments added in every retelling. In the *Metamorphoses* (*c.* AD 2–8), Ovid narrates what appears to be just such a tsunami memory in his section of the poem headed 'Deluge':

> And Neptune with his trident smote the Earth,
> Which trembling with unwonted throes heaved up
> The sources of her waters bare; and through
> Her open plains the rapid rivers rushed
> Resistless, onward bearing the waving grain,
> The budding groves, the houses, sheep and men, –
> And holy temples, and their sacred urns.
> The mansions that remained, resisting vast
> And total ruin, deepening waves concealed
> And whelmed their tottering turrets in the flood
> And whirling gulf. And now one vast expanse,
> The land and sea were mingled in the waste
> Of endless waves – a sea without a shore.[4]

Again, the circumstantial details accord strongly with what we know of earthquake and tsunami behaviour, in spite of the mythic elaborations that surround them. Gods and miracles aside, at heart these are finely observed descriptions of the sea, drawn from the myths and memories of an experienced maritime culture.

Post-classical literature, by contrast, showed little engagement with geophysical events. Even in Japan, the most tsunami-battered nation on earth, surprisingly few pre-modern writers or artists responded to the perpetual seismicity of everyday life. One of the few who did was the twelfth-century literary recluse Kamo no Chōmei, whose influential essay-memoir *Hōjōki* ('The Ten Foot Square Hut', 1212) contains a rare description of a medieval tsunami, offered as a sobering theme for Buddhist contemplation:

> In the second year of the era Gen-ryaku [1132] there was a great earthquake. And this was no ordinary one. The hills

The frontispiece of an early 18th-century edition of Daniel Defoe's *Robinson Crusoe* features stormy seas with mountainous waves battering the island. The novel contains the first tsunami in English literature.

crumbled down and filled the rivers, and the sea surged up and overwhelmed the land. The earth split asunder and water gushed out. The rocks broke off and rolled down into the valleys, while boats at sea staggered in the swell and horses on land could find no sure foothold ... For one terror following on another there is nothing equal to an earthquake.[5]

Aftershocks continued for several weeks, according to Chōmei, for whom the disaster served as a salutary reminder of the impermanence of earthly things; better to live detached from the world, he advised, than to invest any importance in one's physical environment. Life may be tough, but it is fleeting ...

Chōmei spent much of his later life alone in a hut on a hill, and his poems and essays did much to popularize the Japanese genre of *sōan bungaku* ('recluse literature'), of which Daniel Defoe's *Robinson Crusoe* (1719) was a later Western variant. By a strange coincidence the earliest tsunami in English literature can be found in Defoe's great novel. The episode occurs after Crusoe has been marooned on his South American island for several months, and has just finished building a permanent shelter. 'All on a sudden', he writes, 'I found the earth come crumbling down from the roof of my cave':

> I plainly saw it was a terrible earthquake, for the ground I stood on shook three times at about eight minutes distance, with three such shocks as would have overturned the strongest building that could be suppos'd to have stood on the earth . . . I perceiv'd also, the very sea was put into violent motion by it; and I believe the shocks were stronger under the water than on the island.[6]

The tsunami was quickly followed by a hurricane, which Crusoe also attributed to the earthquake, but in every other respect Defoe's geography was sound: the islands off Chile are prone to regular tsunamis, one of which, in February 2010, killed sixteen people on Isla Robinson Crusoe itself (the largest island in the Juan Fernández archipelago, some 600 km (370 miles) west of Chile), following a powerful undersea earthquake. Alexander Selkirk, the marooned Scottish sailor on whom Crusoe was based, is likely to have experienced a tsunami or its aftermath, as he was on the same small island (then named Más a Tierra) in 1705, when an offshore earthquake generated a destructive tsunami that swept across the archipelago and onto the coast of Chile. Whether Selkirk and Defoe ever met is open to question – there are claims that the pair were introduced in a tavern in Bristol – but Selkirk certainly told his story to other London writers, including the journalist Richard Steele, who made Selkirk famous long before he was turned into fiction.

As was seen in an earlier chapter, the Lisbon earthquake traumatized Europe, a continent that had long regarded itself as seismically secure. As the philosopher Susan Neiman observed, 'the eighteenth century used the word *Lisbon* much as we use the word *Auschwitz* today . . . it takes no more than the name of a place to mean: the collapse of the most basic trust in the world, the grounds that make civilization possible'.[7] This deep unease found its most memorable expression in the work of the French satirist Voltaire, who had been provoked by what he saw as complacency expressed in the wake of the disaster. His bleak 240-line *Poème sur le désastre de Lisbonne* (1755), published only a few weeks after the event, invited optimists and moralists alike to gather round and contemplate 'this ruin of a world', his intention being to remind his readers of 'the sad and ancient truth, recognised by all men, that evil walks the earth'.[8] Optimism, he wrote, along with the belief in a benign providence, was not only a foolish but a pernicious response to the pain of human existence, a view that he developed further in his best-known work, *Candide; ou, l'optimisme* (1759), a vitriolic satire that continued from where the Lisbon poem left off.

Like Defoe, Voltaire exposed his hero to an unforgiving sequence of storm, shipwreck, earthquake and tsunami, none of which could dent Candide's overriding conviction that 'all is for the best in the best of all possible worlds'. The novel's bleakest episode takes place in Lisbon on the day of the disaster, as Candide and his travelling companion, the ever-optimistic Dr Pangloss, venture ashore:

> Scarcely had they reached the town when they felt the earth tremble beneath them. The sea boiled up in the harbour and broke the ships which lay at anchor. Whirlwinds of flame and ashes covered the streets and squares. Houses came crashing down. Roofs toppled on to their foundations, and the foundations crumbled . . . 'What can be the "sufficient reason" for this phenomenon?' said Pangloss.[9]

The darkness of Voltaire's vision perturbed many of his contemporaries, and did nothing to reassure later visitors to Lisbon, over which the spectre of destruction continued to hover long after it had been rebuilt. When Charles Dickens visited the city in the mid-1850s, a century after the earthquake, he found himself prey to recurring fantasies of ruin:

I imagined myself, that November morning, on that safe roof-
top watching the tranquil city. Suddenly, the houses all around
me began to roll and tremble like a stormy sea. Through an
eclipse dimness I saw the buildings round my feet and far away
on every side gape, and split; the floors fell with the shake of
cannons. The groans and cries of a great battle were round
me. I could hear the sea dashing on the quays, and rising to
swallow what the earthquake had left . . .[10]

A hundred years on, the event had lost none of its narrative allure, as a barrowload of nineteenth-century novels attest.
From Manuel Pinheiro Chagas's *O Terremoto de Lisboa* (1874),
to Edward Bynner's *Agnes Surriage* (1888), the earthquake and
tsunami provided the dramatic backdrop to any number of
historical romances. Sir Arthur Quiller-Couch's *Lady Good-for-
Nothing* (1910) was a typical production, an elaborate love story that
culminated in Lisbon on the day of the disaster. Quiller-Couch's
description of the tsunami, which was evidently based on eyewitness accounts, was the highlight of an otherwise forgettable
novel:

> She stared. She did not comprehend; she only saw that a
> stroke more awful than any was falling, or about to fall. The
> first convulsion had lifted the river bed, leaving the anchored
> ships high and dry. Some lay canted almost on their beam
> ends. As the bottom sank again they slowly righted, but too
> late; for the mass of water, flung to the opposite shore, and
> hurled back from it, came swooping with a refluent wave,
> that even from this high hillside was seen to be monstrous.
> It fell on their decks, drowning and smothering: their masts
> only were visible above the smother, some pointing firmly,
> others tottering and breaking . . . and still, its crest arching,
> its deep note gathering, the great wave came on straight for
> the harbour quay.

Just as the wave was about to hit land, it was obscured from view
by a cloud of smoke (perhaps Sir Arthur was losing the will to

describe), although the sound it made was terrible, 'drowning the last cry of thousands':

> for before it died away earthquake and wave together had turned the harbour quay of Lisbon bottom up, and engulfed it. Of all the population huddled there to escape from death in the falling streets, not a corpse ever rose to the surface of the Tagus.[11]

The eruption of Krakatoa spawned a similar legacy of 'good bad' novels, in which the tsunami vied with the exploding volcano for sheer destructive spectacle. R. M. Ballantyne's *Blown to Bits; or, the Lonely Man of Rakata* (1889) interwove a far-fetched tale of exploration and revenge with a detailed, factually based account of Krakatoa's final act. The novel's central dramatic set-piece sees its hero, Nigel Roy, facing down the wave in his two-masted brig as the vessel zigzags through the Sunda Strait in the shadow of the erupting volcano:

> At each of these explosions a tremendous sea-wave was created by the volcano, which swept like a watery ring from Krakatoa as a centre to the surrounding shores . . . there, rushing on with awful speed and a sort of hissing roar, came the monstrous wave, emerging, as it were, out of thick darkness, like a mighty wall of water with a foaming white crest, not much less – according to an average of the most reliable estimates – than 100 feet high.[12]

A sudden lava-flare from the volcano reveals the terrifying fact that the ship is no longer above the seabed, but being carried 'right through, or rather *over*, the demolished town of Anjer', heading far inland with the turbulent wave, in a fictional rendering of the journey of the *Berouw*, whose fate had come to stand as a ready image of the power and ferocity of the sea.

Disaster novels invariably contain passages of technical exposition, often in the midst of the action, as a signal to the reader that the fiction is rooted in fact. *Blown to Bits* features an unusually detailed explanatory section, drawn from the Royal Society's wide-ranging report, *The Eruption of Krakatoa, and Subsequent Phenomena* (1888), in which the testimonies of 'some of the best-known men of science' were woven into the narrative:

> Mr Verbeek, in his treatise on this subject, estimates that a cubic mile of Krakatoa was propelled in the form of the finest dust into the higher regions of the atmosphere – probably about thirty miles! . . . and the mighty waves which caused such destruction in the vicinity of Sunda Straits travelled not once, but at least six times round the globe, as was proved by trustworthy and independent observations of tide-gauges and barometers made and recorded at the same time in nearly all lands – including our own.[13]

In a similar vein, H. E. Raabe's *Krakatoa: Hand of the Gods* (1930) saw an implausible and, at times, incoherent storyline enlivened by a number of factual asides sourced from a layperson's guide to the sea. The result was a curious tonal mix of believability and extravagance:

> The crest was toppling. A matter of moments now and I would be crushed, dashed against the side of some hill, later to be left there by the receding waters . . . But I neither sank nor dropped through space. Instead I felt myself gliding downward, like on a toboggan descending on a sloping plain. The enormous bulk of water below, having swept away all of the ground-resistance, was now advancing faster than the top could follow and the curling crest was drawn back into the flattening slope . . . Those who have watched the breakers at the seashore can often see the highest of them pull down their own white-caps, crush down the smaller ones ahead of them and sweep over the sand more like a rolling swell. That law, governing the waves, was what saved me then. But no wave like that one had ever been watched by man at the shore.[14]

As seismology developed over the twentieth century, so did its role in fiction. In what is still the best-known tsunami novel, Paul Gallico's *The Poseidon Adventure* (1969), recent advances in undersea geology were recruited to explain how 'an unaccountable swell, 400 miles south-west of the Azores' succeeds in overturning the legendary cruise ship as it steams away from Lisbon harbour one fateful Boxing Day morning. At one point the captain is handed a message broadcast from a nearby seismographic station informing him of 'a mild, rolling, sub-sea earthquake of no great duration resulting in the build-up of the swell affecting vessels to the south', making him the first character in literature to receive an automated tsunami warning. More impressively, the existence of the Mid-Atlantic Ridge had only been confirmed by oceanographers a year or so before the novel appeared, but here it is playing the inciting role in what became (in the wake of the 1972 film version) the greatest disaster story of its time:

> [As the ship passed over] the huge fault known to exist in the Ridge at exactly eight minutes past nine, this fault already weakened by the preliminary tremor, now without warning shifted violently and slipped a hundred or so feet, sucking down with it some billions of tons of water . . . when the s.s. *Poseidon* met the gigantic, upcurling, seismic wave created by the rock slip, she was more than three-quarters broadside, heeling further from the turn. Top heavy and out of trim, she did not even hang for an instant at the point of no return, but was rolled over, bottom up, as swiftly and easily as an eight-hundred ton trawler in a North Atlantic storm.[15]

The Poseidon Adventure was distinguished by the seamless integration of its technical and dramatic elements, as well as by an unusual plot structure that inverted the established disaster convention of a protagonist drawn into a race against time – a trope that has spawned a shelf's worth of near-identical American tsunami thrillers, including Richard Martin Stern's

Tsunami (1988), in which a killer wave is precipitated by undersea weapons testing; Crawford Kilian's *Tsunami* (1999), in which the United States is threatened by a tsunami from Antarctica; Michael Crichton's *State of Fear* (2004), in which environmental terrorists plot an artificial tsunami; Gordon Gumpertz's *Tsunami* (2008), in which an undersea eruption generates a Pacific-wide wave, and J. G. Sandom's *The Wave* (2010), in which terrorists, again, are foiled in their plot to unleash a transatlantic tsunami.

Perhaps the most interesting of these 'scientific' tsunami thrillers is Boyd Morrison's *Rogue Wave* (2010), in which an asteroid impact in the central Pacific launches a megatsunami directly towards Hawaii. The novel's hero, Kai Tanaka, has just been appointed director of the Pacific Tsunami Warning Center, from where he is able to monitor the advance of the wave in 'real' narrative time ('53 minutes to impact', etc.). His dialogue, inevitably, reads like a textbook, which is presumably where much of the technical information came from:

> 'We know how big the earthquake was and how deep the water is in that part of the ocean,' Kai said. 'They developed a formula that would approximate the magnitude of the resulting earthquake depending on the size of the asteroid. We'll solve the formula in reverse based on the size of the quake. From that, we can estimate how big the waves would be at various distances from the impact zone.'[16]

In contrast to such science-laden thrillers, Japanese disaster narratives have tended to invoke wholesale, even global, destruction, with little in the way of detailed exposition – apocalyptic fantasies being a well-established theme of post-war Japanese fiction. Nobel laureate Kenzaburo Ōe's 1973 novel *Kōzui wa waga tamashii ni oyobi* ('The Floodwaters Have Come unto My Soul'), explores a quasi-Biblical disaster (the title is taken from Psalm 69: 'Save me, O God, for the waters are come in unto my soul') in which the world is destroyed by a nuclear accident, followed by a vast wave: almost a prophecy in reverse of what happened in 2011. Ōe's protagonist, sheltered in his underground bunker, imagines 'a huge movement in the earth's crust and the land being visited by tidal waves or floods ... perhaps this great water will wipe out the human race'.[17] A similar strain of apocalyptic nihilism can be found in Sakyo Komatsu's 1973 disaster novel *Nippon chinbotsu* ('Japan Sinks'), another prophetic fantasy of national submersion, in which Japan falls victim to its own geophysics – volcanoes, earthquakes and 30-ft tsunamis. Unlike some of the film versions (see below), in which the country is saved at the eleventh hour, the novel itself ends, shockingly, with the complete physical erasure of Japan:

> In northeastern Honshu the coastline had slid some sixty feet into the Pacific. The waters were rushing in on Hokkaido, too, and sections of it had fallen away ... While the Pacific coast slid away into the deep, the Japan Sea coast rose up for a brief moment, like one side of a capsizing vessel. But then the same blind force took hold of it and plunged it, too, down into the sea.[18]

As the cultural historian Susan J. Napier has observed, this vision of Japan's annihilation was delivered without sensationalism or excitement, but rather with 'a sense of mourning for the loss of Japanese culture'.[19] For Komatsu's generation, numbed and demoralized by the disastrous war and its atomic conclusion, the traditional Japanese preoccupation with ephemerality could be extended from the ubiquitous cherry blossom to the nation itself, a great if fragile culture destroyed by an irresistible force.

Such end-of-the-world scenarios have also been a long-term staple of manga and anime (Japanese comics and cartoons). Kaiji Kawaguchi's *A Spirit of the Sun* (2003–8) was a popular manga series depicting life in a post-apocalyptic Japan torn apart by earthquakes and tsunamis. Under the pretence of providing humanitarian aid, the Chinese army eventually takes over what becomes known as North Japan, while the u.s. military occupies the south, in a political refashioning of the 'sinking of Japan' motif. In the months following the 2011 tsunami, many leading *mangaka* (comic artists) created work in response to the disaster, often as a means to raise money for emergency aid. Koji Yoshimoto's *Santetsu* (2012) told the story of the rebuilding of the Sanriku railway line following its destruction by the tsunami, while Suzuki Miso's *The Day Japan and I Shook* (2011–12) took a more personal approach to the trauma of '3/11', the author interrupting his narrative to air doubts about the ethical propriety of drawing the killer wave:

> Scenes of the tsunami swallowing up cars and houses criss-crosses my mind. With so many people having lost their homes, their families, their land, living in refugee camps and in fear of radiation, what kind of manga should one draw? Is it the time even to be drawing manga?[20]

It's a question that neither he nor his narrator can easily answer – though when the celebrated anime director Hayao Miyazaki was asked why his film *Ponyo* (2008) featured a benign blue tsunami that wet people's feet but did not kill, his thoughtful

Still from the Japanese anime *Ponyo* (dir. Hayao Miyazaki, 2008), which features a 'magical' tsunami.

reply could well be borrowed as an answer to Suzuki's question: 'There is no point in portraying these natural disasters as evil events. They are one of the givens of the world in which we live.'[21]

Tsunamis in visual art

One of the most popular images among Pacific island surfers is a painting by the Hawaiian artist Herb Kane, showing the legendary exploit of Holoua, who claimed to have survived the 1868 Big Island tsunami by surfing to safety on a door panel. According to the story, Holoua and his wife were at home on the afternoon of 2 April, when they felt an unusually strong earthquake. They immediately ran outside towards higher ground, until Holoua decided to head back to the house to retrieve some money that had been left behind:

> Just as he entered the house the sea broke on the shore, and enveloped the building; first washing it several yards inland, and then, as the wave receded, swept it off to sea, with him in it. Being a powerful man, and one of the most expert swimmers in that region, he succeeded in wrenching off a board or a rafter, and with this as a *papa hee nalu* (surfboard) he boldly struck out for the shore, and landed safely with the return wave.[22]

Considering that the wave was reputed to have been 18 m (60 ft) high, and filled with churning debris, the tale seems scarcely credible, though it was reported as fact in the *Hawaiian Gazette* for 29 April 1868, and has lingered in Pacific surfing lore ever since. Wiry Holoua, with his wallet between his teeth and his hands guiding the board to shore, remains a compelling figure in the history of the sport, though his story reinforces a reckless outlook exemplified by the popular motto: 'In case of earthquake, go get your surfboard.'

There is, of course, no shortage of eyewitness testimony to confirm that a Pacific tsunami is not, in reality, a clean, surfable

wave but a dangerous, debris-laden flood that bulldozes its way inland. Masuo Kino, a survivor of the 1946 April Fool's Day tsunami, recalled the incoming wave as

C Katsushika Hokusai, 'The Great Wave off Kanagawa', from the series *Thirty-Six Views of Mount Fuji* (*c.* 1830).

> a huge wall of water. And it wasn't a beautiful wave like you see in surfing magazines. It was just a wall of gray, black water. And as we just stood there and watched it got bigger and bigger and closer and closer . . . and then we ran.[23]

Yet the myth of a tsunami as a cresting blue breaker is worryingly persistent, reinforced in part by the ubiquity of one of the world's most reproduced images, Katsushika Hokusai's *Great Wave off Kanagawa* (*c.* 1830).

Hokusai's woodblock print is one of the most celebrated icons of Japanese art. It shows a towering storm wave about to engulf three fishing vessels in the waters off the southern Kantō region of Japan. In the distance, a vulnerable-looking Mount Fuji, the symbol of Japan, can be glimpsed through the trough of the breaker, like a scene from the end of *Nippon chinbotsu*. The image has attracted a wealth of scholarly interest over the years, some of it concerned with the physical nature of the wave itself, which, while clearly not a tsunami, has often been described as one. An impassioned article by the Hawaiian seismologist Doak C. Cox lamented 'the erroneous identification of the "Great Wave" as a tsunami wave and its inappropriate use as a tsunami icon', arguing that the misidentification of Hokusai's storm wave gives a false and potentially dangerous impression of actual tsunami behaviour.[24] Cox was particularly dismayed by the widespread use of the image by organizations such as the International Tsunami Information Center, which uses a silhouette of the famous wave on its *Tsunami Newsletter*, as does the Pacific Tsunami Museum in Hilo, Hawaii. The United States Geological Survey also features the image on its 'Tsunamis & Earthquakes' webpage, while coastguard agencies around the world have adopted a simplified version for use on tsunami hazard zone signage. No wonder tsunamis remain so strongly associated with Hokusai's plunging breaker.

The *Great Wave*'s influence goes back a long way. One of the nineteenth century's most reproduced tsunami images was

Tsunami Hazard Zone warning sign, an example of the pervasive design influence of Hokusai's famous wave.

a newspaper engraving of the Royal Mail Steamship *La Plata* riding up the face of a giant wave in the eastern Caribbean. Yet the smoking volcano seen in the distance recalls Hokusai's iconic Mount Fuji, while the perilous drama of the composition – will the steamship and its two coal sculls survive the great wave? – echoes his motif of a trio of boats in peril. The image was made to accompany an extraordinary eyewitness testimony that was published in many of the world's newspapers and magazines. The account's author, a young New Zealander named William Miles Maskell, had been a paying passenger on *La Plata*, which on the serene afternoon of 18 November 1867 was taking on coal some 4 km (2½ miles) southwest of St Thomas Harbour, a busy port in what were then the Danish Virgin Islands. At around 2.20 p.m. a pair of violent undersea earthquakes shook the vessel, their epicentre apparently close by. Fifteen minutes later, according to Maskell, 'like a gigantic wall, the tremendous earthquake wave was rushing towards us'. Those on board the steamship had less than five minutes to prepare themselves before 'the terrible mountain of water' was upon them:

> Down on the devoted ship came the roaring sea, literally piled up like a wall, rolling upon us at the rate of fifty miles an hour, with a perpendicular face (as was afterwards ascertained, by

actual measurement, at one of the beacons) of fully forty feet ... Most fortunately for us, about half a mile outside of our position, Water Island ended in a low point. This, exposed to the full force of the wave, met it like a wedge; and it was observed that the moment the sea reached this point it broke, and dropped, so that when it struck the ship, it was perhaps not more than ten feet higher than the taffrail. We had been lying all the morning head to sea; but the water receding from the shore like a mill race to meet the wave, turned the ship round, so that it took us on the stern. When the first rush of water came it slewed her a little more, and the main force of the wave itself fell full on the starboard quarter. Three waves in one it came, with a roar as of a hundred Niagaras; and contrary to all expectation, the *La Plata* rose over it like a duck, and was saved![25]

Film poster for *The Perfect Storm* (dir. Wolfgang Petersen, 2000), testament to the long-term influence of the *La Plata* engraving.

The wave then crashed into St Thomas Harbour, demolishing dozens of ships, and sweeping many people into the water. Other islands, notably St John to the east and St Croix to the south, were also badly hit. In all 23 people were killed in the Virgin Islands tsunami, a comparatively minor disaster that might well have been forgotten, were it not for the totemic image that it gifted to the world: a solitary vessel being driven up the face of a mountainous wave, an image that has reappeared in many other contexts, including in the special effects for films such as *The Poseidon Adventure* (1972) and *The Perfect Storm* (2000).

Variations on these nineteenth-century precedents are still being produced today. One dramatic example by the British space artist David A. Hardy features another unmistakable reference to Hokusai's distant view of Mount Fuji, though the fishing boats have been replaced by a submerged temple and trees. Lynette Cook's stylized rendering of the Krakatoa tsunami, meanwhile, features a doomed fishing boat rising up the face of the wave in apparent homage to the *La Plata* image.

Yet Hokusai's wave was an untypical subject for the art of Edo-period Japan (1600–1867), which was characterized instead by the sensual *ukiyo* ('Floating World') culture that consciously turned its back on the depiction of earthly affairs. There are consequently few images of earthquakes and tsunamis among the celebrated canon of Japanese prints known as *ukiyo-e*, or 'pictures of the Floating World'. An abundance of geishas, tea-houses and shrines also characterizes Edo-era literature, but there is little in the way of the nation's turbulent geology. One rare example was a view of Mount Fuji by Utagawa Hiroshige, which featured an image of the small crater (known as Mount Hōei) that was formed during Fuji's last eruption in December 1707, while the handful of Edo-era images of tsunami waves appeared as book illustrations depicting the Ansei-Nankai earthquake and tsunami of December 1854.

What did arise from the *ukiyo-e* tradition, particularly towards the end of the Edo period, were the prints known as *namazu-e*, propitiatory images depicting the giant subterranean catfish Namazu who, as was seen in the previous chapter, was held to be

David A. Hardy's 'Tsunami', from *The Fires Within: Volcanoes on Earth & Other Planets* (1991), features a knowing reference to Hokusai's Mount Fuji.

The doomed fishing boat in Lynette Cook's artwork of a tsunami advancing on the port of Anjer following the Krakatoa eruption of 1883 seems to echo both Hokusai's 'Great Wave' and the later *La Plata* image.

responsible for earthquakes and tsunamis. Namazu prints and poems had been popular since the seventeenth century. A linked verse published in 1678 by the poet Matsuo Basho begins by asking whether earthquakes are caused by a dragon writhing in the underworld, and answers: 'No, they are a giant catfish moving' (*'Takejujo-no namazu narikeri'*).[26]

The giant catfish Namazu being attacked by angry peasants following the 1855 Ansei-Edo earthquake. A woodblock print of 1855.

But it was the Ansei-Edo earthquake of 11 November 1855 that gave rise to the greatest outpouring of such images, a pheno - menon that has been interpreted as part of a wider protest against the unpopular government of the time. As the historian Gregory Smits has written, natural disasters often act as cultural catalysts, and the Ansei earthquake 'shook the social and political foundations of Edo along with the earth's crust, and the *namazu-e* were the reaction of the common people to this event'.[27] Many of the 1855 *namazu-e* show the people attacking Namazu, taking over the god Kashima's official function, in an act of symbolic resistance to arbitrary authority. This visual tradition continues

Zuza Miśko's 'That's enough, Namazu' depicts the Slavic demon Rusalka restraining Namazu with her braid. Note the catfish's atomic eyes, a reference to the Fukushima meltdown. The print was made to raise money for the victims of the 2011 disaster.

today, with catfish pictures once more circulating throughout an anxious Japan in the wake of the Fukushima disaster.

It wasn't until the 1896 Meiji Sanriku tsunami (nearly 30 years after the end of the isolationist Edo period) that images of Japanese tsunamis began to appear in any number. But the briefest glance at a proto-expressionist painting such as Shoukoku Yamamoto's *Houses, Men and Cattle are Washed Away* (1896) makes it clear that the decorous *ukiyo-e* era was well and truly over, and that Japan was no longer floating, it was sinking.

This *Vue de l'Optique* composition (a hand-coloured copper engraving used in a magic lantern) shows ships and boats being tossed about in the Messina Strait by the 1783 Calabria earthquake and tsunami.

In Europe, meanwhile, earthquakes and 'tidal waves' had been long-established subjects for popular culture. A number of surviving magic lantern slides feature violent tempests and mountainous waves, with one particularly beautiful late-eighteenth-century engraving showing ships and boats being hurled about in the Messina Strait by the Calabrian earthquake and tsunami of February 1783.

The Lisbon and Calabrian earthquakes had coincided with the development of a range of innovative visual technologies such as the panorama, the cyclorama and the Eidophusikon: pre-electric forerunners of the modern cinema, for which the drama of an earthquake proved an enduring subject. In one version,

Tsunamis in popular culture: this Wills's cigarette card from 1924 shows a large 'tidal wave' striking a tropical coastline. The caption reads: 'After the earthquake which destroyed Messina, Sicily, a tidal wave devastates the coast of South America!', a claim that appears to conflate (and reverse) the tsunamis of 1906 and 1908 (see the Timeline).

WILLS'S CIGARETTES.

A TIDAL WAVE.

Houses, Men and Cattle are Washed Away, a painting of 1896 by Shoukoku Yamamoto, depicting the Meiji Sanriku tsunami.

'The Cyclorama of Lisbon', which went on show in a specially constructed circular theatre in 1848, the destruction of the city was shown through a sequence of revolving scenes, accompanied by live organ music including, inevitably, Haydn's *Il Terremoto* ('the earthquake'). An enthusiastic review in the *Spectator* gave a detailed description of the 'novel and strikingly beautiful exhibition', although the unsigned author evidently mistook the depiction of the tsunami for that of a sea storm:

A 19th-century Japanese print illustrating the Meiji Sanriku earthquake of 15 June 1896.

When the grand square of Lisbon is reached, the scene is darkened; rumbling and rushing sounds are heard, followed by the tolling of bells and the crash of falling buildings,– conveying as good an idea of the horrors of an earthquake as mechanical means may produce. The scene now changes to the sea, which is surging extraordinarily, and vessels are tossing about in all directions: altogether a very effective representation in miniature of a storm at sea.[28]

The painted tsunami scene from 'The Cyclorama of Lisbon', which went on show in 1848 in a specially constructed circular theatre at the London Colosseum.

But such sound and vision spectaculars were only a foretaste of things to come in the following century, when the disaster movie arrived to take centre stage in mass-market entertainment.

Tsunamis in film

The advent of cinema in the late nineteenth century saw the representation of natural disasters reach yet more spectacular heights. The first big-budget tsunami picture was RKO's apocalyptic *Deluge* (dir. Felix E. Feist, 1933), which featured an impressive special-effects tsunami bearing down on an earthquake-ravaged New York. The much-admired wave sequence,

Film poster for *Deluge* (dir. Felix E. Feist, 1933), the first big-budget tsunami movie, in which a giant wave annihilates New York City.

Still from the much-reused wave sequence of *Deluge* (1933).

created by the cinematographer William Williams, was reused in a number of later films, including *S.O.S. Tidal Wave* (dir. John H. Auer, 1939), *Dick Tracy vs Crime Inc.* (dir. John English, 1941) and, spectacularly, in the dramatic final episode of the Republic Pictures series *King of the Rocket Men* (dir. Fred C. Brannon, 1949), in which the 'worst earthquake and tidal wave in American history' is generated by the villainous Dr Vulcan, using a stolen thorium-wave 'Decimator'. Though the latter's conceit of the anthropogenic natural disaster became a staple of post-war science fiction, the majority of mainstream disaster movies continued to pit mankind firmly against nature.

The brazenly mistitled *Krakatoa, East of Java* (dir. Bernard Kowalski, 1969) set the tone for the 1970s golden age of disaster movies. Filmed in 70 mm Super Panavision, through which the erupting volcano and its accompanying tsunami blazed from the screen 'like an animated John Martin painting', in the words of *The Times*'s film critic, the Oscar-nominated special effects raised the bar for the spectaculars that followed, though the film had to be retitled *Volcano* in response to near-universal derision (Krakatoa, of course, being *west* of Java).[29]

Paul Gallico's novel provided the source material for the twentieth century's most enduring disaster movie, *The Poseidon Adventure* (dir. Ronald Neame, 1972), in which a giant wave kills off more than a thousand nameless extras in the first few minutes, leaving only a handful of survivors (led by Gene Hackman) to battle their way to safety. As film critic Stephen Keane observed, the script leant heavily on the novel's implied Greek/Christian symbolism, with the captain's ruminations on the vessel's name – 'Poseidon: in Greek mythology the god of the sea, also god of storms, tempests, earthquakes and other natural disasters. Quite an ill-tempered fellow' – countered by a near-constant litany of Christian oaths:

> With the main tidal wave approaching, the captain can only say: 'Oh my God!', and similarly, after the wave has hit, one

Poster from *The Poseidon Adventure* (dir. Ronald Neame, 1972).

The Hokusai-
influenced film poster
for *The Last Wave*
(dir. Peter Weir, 1977).

of the passengers can only ask: 'Jesus Christ, what happened?' *The Poseidon Adventure* is packed with people taking the Lord's name in vain.[30]

Much of the film was spent exploring the complex moralities of sacrifice and survival, themes that have preoccupied disaster movies ever since, from *The Towering Inferno* (1974) to *The Impossible* (2012), with Peter Weir's *The Last Wave* (1977) seeking to broaden the moral canvas from the personal to the eco-political; the film's closing shot is filled by a vast, punitive wave that crashes over Sydney in apparent planetary punishment of modern Australia's continued mistreatment of both its environment and its Aboriginal peoples.

Ever since the first *Godzilla* picture in 1954, Japan has been a notably prolific producer and consumer of disaster films. Sakyo

Promotional poster for *Tidal Wave* (dir. Andrew Meyer, 1974), a straight-to-video American remake of the popular Japanese disaster movie *Nippon chinbotsu* (*Japan Sinks*, 1973).

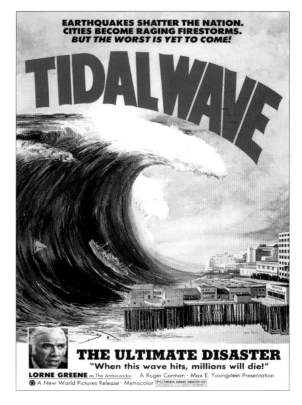

Komatsu's novel *Nippon chinbotsu* ('Japan Sinks'), discussed above, has been adapted many times for cinema and television. The first film version, *Nippon chinbotsu* (dir. Shirô Moritani, 1973), was a four-hour epic that proved a popular domestic hit. Remaining faithful to the novel's bleak vision (unlike some of the more recent remakes), it ends with the obliteration of Japan, confirmed by a lingering aerial shot over the sea, 'the names of its no-longer-existing cities superimposed on an empty ocean. All that remains of Japan is its history', notes Susan J. Napier, who reads this fantasy of national submersion as an homage to Japan's waning cultural identity.[31] The film's popularity at home stands in revealing contrast to the fortunes of the edited English-language version, *Tidal Wave* (dir. Andrew Meyer, 1974), which went straight to video in Europe and America, never to be heard of again.

Thirty years later, Komatsu's novel was filmed again as *The Sinking of Japan* (dir. Shinji Higuchi, 2006), having in the meantime fuelled any number of best-selling mangas and television animes – indeed, the 'sinking' trope had become so embedded in Japanese popular culture that a parody movie entitled *Nippon Igai Zenbu Chinbotsu* ('Everything Other than Japan Sinks') was also released in 2006, directed by Minoru Kawasaki, in which every landmass on the planet is submerged by violent tectonic upheaval, leaving only the Japanese islands intact. Ironically, the film was set in 2011, the year of the tsunami, when a long section of Japan's northeast coastline did in fact subside into the sea, allowing the waves to penetrate several kilometres inland.

News footage of the Japanese and Indian Ocean tsunamis is now among the most familiar imagery of modern times, so it is hard to recall how few moving images of tsunamis existed before the turn of the twenty-first century. When the computer graphics artists who worked on Mimi Leder's disaster movie *Deep Impact* (1998) were tasked with simulating a cosmogenic megatsunami they could find no suitable clips to work from: 'I imagine anyone trying to film a tidal wave would be dead', as one of the designers commented in a magazine interview, and

Film still from *Deep Impact* (1998).

much of the film's celebrated wave sequence was created from a combination of surf footage, physics textbooks and visual experimentation.[32] The results, however, were impressive, and *Deep Impact* remains a rarity, a Hollywood disaster film in which the spectacle and the science were equally credible.

While high-grossing films such as *Deep Impact* and Roland Emmerich's *The Day After Tomorrow* (2004) set out to explore emerging anxieties around global catastrophe and meltdown, the Boxing Day tsunami supplied a shocking real-world example, an instant symbol of apparent planetary collapse. One of the most powerful cinematic responses to the scale of the disaster was the Tamil-language science fiction epic *Dasavathaaram* (dir. K. S. Ravikumar, 2008), which was set in a variety of time frames from the twelfth to the twenty-first centuries, leading up to the morning of the 2004 tsunami. The film, which blended ancient

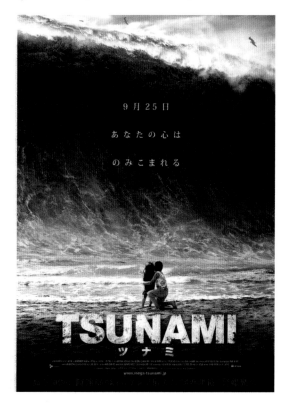

Promotional poster for *Haeundae* (dir. Yoon Je-kyoon, 2009), South Korea's first big-budget disaster movie, released in an English-language version as *Tidal Wave*, and in Japan as *Tsunami*.

Film poster for *The Impossible* (dir. Juan Antonio Bayona, 2012), based on the true story of a family on holiday in Thailand in December 2004.

THE IMPOSSIBLE

myth and prophecy with a convoluted sci-fi plot, broke box-office records across Asia, where it was seen by more than 20 million cinema-goers. Two years later, Clint Eastwood's equally ambitious, multi-layered disaster epic *Hereafter* (2010), had only just been released in Japan when the 2011 tsunami struck. The film, which opens with a powerfully realistic rendition of the Boxing Day tsunami (the special effects were justifiably nominated for an Oscar), was withdrawn from all Japanese cinemas in response to the surrounding trauma. Such gestures run deeper than simple respect for the victims: as Susan Sontag observed in her influential essay, 'The Imagination of Disaster' (1965), disaster movies are framed around a peculiarly modern anxiety, the fear of 'collective incineration and extinction which could come at any time, virtually without warning'; so when the real thing does arrive, as happened in December 2004 and

March 2011, its representation becomes immediately and collectively unbearable.[33] As Suzuki Miso discovered in the wake of the Japan tsunami ('is it the time even to be drawing manga?'), a certain amount of time needs to pass before an event can be transformed into a memory.

There are, however, exceptions: Lucy Walker's Oscar-nominated documentary, *The Tsunami and the Cherry Blossom* (2011), featured interviews with Japanese tsunami survivors filmed amid the onset of cherry blossom season, exactly one month after the disaster. As Walker's interviewees told their stories, falling blossom drifted over the carnage. 'Nature has a dreadful destructive power, and nature has a positive creative power', as one survivor observed. 'There is always beauty and terror in nature, but we forgot the terror.' And he was right: our nemesis is not nature, but amnesia. Many of those who died in Japan had put their faith in sea walls and sirens, and had waited for the wave to be repelled: 'Maybe like the one last year – only a few feet high. The floodgate will protect us for sure.'[34] Such willed forgetting is a near-universal human trait and, as the next and final chapter will suggest, the first step to protecting ourselves from future disasters is to learn how to keep in mind the lessons of the past.

It is, of course, far easier said than done.

5 Living with Tsunamis: Warning Systems and Coastal Defence

With less than an hour of warning and a simple lesson in advance on what to do, most would have been able to simply walk a mile inland to safety and the death toll would have been counted in the hundreds rather than the tens of thousands. Providing these things is not advanced science.[1]

On 23 December 1854 readings retrieved from a set of automated tide gauges off the coast of California displayed some unusual oscillations, but as there had been no storms or observed tidal anomalies in the area, hydrographers were puzzled. It couldn't have been their new equipment, which was state of the art and highly reliable, consisting of a series of floats attached to slowly rotating drums of paper on which continual tide measurements were recorded. So it must have been something in the water.

It took several months for the U.S. Coast Survey to deduce that the oscillations were in fact the signal of a long-range tsunami that had been generated by the Ansei-Nankai earthquake in southeastern Japan. A section of the Japanese coast had been badly hit by the tsunami which, it transpired, had also travelled east across the Pacific. By the time it reached California, the tsunami had weakened considerably, but the waves still showed up as 15-cm (6-in.) zigzags on the tide gauges. But these insignificant-looking scribbles turned out to be a small piece of history, for they were the first tsunami measurements ever taken by an instrument, as well as the first remote sensing of an earthquake via its secondary sea-wave. The potential implications of this discovery were not lost on oceanographers, and over the course of the following century these early traces gave rise to what became the world's first automated tsunami warning system.

Warning systems

The first successful deployment of a modern warning system was in March 1933, although the idea had been in circulation since the self-registering seismograph came into widespread use at the end of the nineteenth century. The road to this new technology, however, was not always smooth.

One February morning in 1923, Thomas A. Jaggar, director of the Hawaiian Volcano Observatory on Kilauea, arrived for work at 8 a.m., and noticed a series of spikes on the office's seismographic printout; these confirmed that a powerful earthquake, somewhere to the north, had occurred three hours earlier. Based on the size and location of the tremor, as well as on his understanding of the behaviour of the sea, Jaggar concluded that a trans-Pacific tsunami was likely to have been generated, and he telephoned the Hilo harbourmaster to warn of a series of big waves that were likely to be on their way. But to Jaggar's incredulity the warning was dismissed, and when the waves began to batter Hawaii shortly after noon, seven hours after the earthquake itself, no precautions had been taken against the incoming sea. The biggest wave, some 6 m (20 ft) in height, destroyed most of the Hilo tuna fleet, killing one of the fishermen, and wrecking dozens of buildings.

Jaggar and his colleagues were understandably downhearted, for they had known for some time that while earthquakes are unpredictable, long-range tsunamis are not. Given that the time in hours taken by tsunamis to cross the Pacific is roughly equal to the time in minutes that the faster-moving seismic shock waves take to cover the same distance, a 60-fold window of warning will usually be available. By using the minutes for hours rule, as one of Jaggar's colleagues pointed out in an article published in 1924, 'it should be possible to predict the arrival time of tsunami waves from all parts of the Pacific', though since the majority of seismographs were only inspected once a day, he suggested that they be fitted with some kind of alarm bell to warn of bigger quakes.[2] The concept of automated tsunami warnings had thus been introduced, and the idea was actually

put into practice ten years later, in March 1933, when a magnitude 8.6 earthquake off the northeast coast of Japan sent 23-m (75-ft) waves crashing into the Sanriku coast, killing more than 3,000 people and wrecking more than 8,000 boats. That was the tsunami that destroyed the village of Aneyoshi, and prompted the installation of its warning stone: 'remember the calamity of the great tsunamis. Do not build any homes below this point'. It also prompted the building of the vast concrete sea walls that today line 40 per cent of the Japanese coast.

The 10-metre-high 'great wall' of Tarou, Iwate Prefecture, Japan; built after the 1933 tsunami, but fatally overwhelmed by the tsunami of March 2011.

The seismograph at Jaggar's observatory had registered the earthquake at 7.10 a.m. Hawaiian time, with analysis of the printouts indicating an undersea epicentre off northeast Japan, some 6,350 km (3,950 miles) away. Just as they had done ten years before, the observatory notified the Hawaiian port authorities that a tsunami was likely to arrive later that day, and this time the warning was heeded. The fishing fleet was moved into Hilo Bay, while on the other side of Hawaii's Big Island, which directly faces distant Japan, stevedores began removing cargo from the docks where, at 3.20 p.m. – eight hours on from the earthquake itself – the sea withdrew in the characteristic *kai mimiki* pattern, before the first of the waves rushed in, causing serious flooding along the coast. Fifteen minutes later, the waves reached Hilo, although, since it faces away from Japan, these 'wraparound' waves were much lower and weaker, causing little in the way of damage.

But the important fact was that the tsunami warning had worked, thanks to the seismographs at the Volcano Observatory, and that this time no one on the Hawaiian islands was killed or injured by the wave. The era of advance tsunami warnings had arrived. But reliance on seismographic readouts alone soon led to warnings being given out every time a major earthquake was recorded. The years following the successful 1933 alert were punctuated by a series of disruptive evacuations of Hilo harbour based on readings from far-off earthquakes in various locations around the Pacific, none of which generated a tsunami. In fact no perceptible tsunami reached Hawaii for more than a decade – not until the disastrous morning of 1 April 1946. By then, the high incidence of 'false alarms' had seen tsunami warnings and evacuations abandoned as an expensive waste of time, while the war years had introduced greater threats to life than tsunamis from distant tremors. 'Less than one in one hundred earthquakes result in tidal waves and you don't alert every port in the Pacific each time a quake occurs', as the much-criticized director of the u.s. Coast and Geodetic Survey pointed out in the wake of the 1946 disaster.[3] But it was clear that a new and more reliable Pacific warning system had long been overdue.

The American oceanographer Francis Shepard had witnessed (and even photographed) the 1946 tsunami at close proximity and lived to tell the tale. He knew that the technology needed to revive the Hawaiian warning system was already available at little cost, and in the paper published in *Pacific Science* in 1947, he and his colleagues laid out a series of recom - mendations for a coordinated, Pacific-wide tsunami warning system:

A system of stations could be established around the shores of the Pacific and on mid-Pacific islands, which would observe either visually or instrumentally the arrival of long waves of the periods characterising tsunamis. The arrival of these waves should be reported immediately to a central station, whose duty it would be to correlate the reports and issue warnings to places in the path of the waves.[4]

These recommendations were taken seriously, and over the coming months an outline warning system began to be drawn up, along with approximate wave transit times to Hawaii from various tsunamigenic hotspots around the Pacific, as drawn from Jaggar's hours-to-minutes rule: five hours from the Aleutian Islands; six hours from Kamchatka; ten hours from Japan; fifteen hours from South America.

By the end of 1948, the Seismic Sea Wave Warning System (it was later renamed the Tsunami Warning System) was up and running in the form of a collection of seismic and tidal information exchanges located on United States-owned Pacific stations including Alaska, Hawaii and Midway Island. Its mechanism was straightforward: any earthquake of magnitude 7.0 or above occurring anywhere in the Pacific (or 6.8 if the focus of the quake was located within the Hawaiian chain) would set off alarms at the monitoring stations, alerting staff to the event. If the seismologists judged that a tsunami was a likely consequence – a judgement based on the quake's location as well as on its magnitude – a tsunami *watch* would be declared, and all relevant civil defence authorities placed on high alert. This was the first, seismographic, stage of any warning cycle. The second, ocean - ographic, stage would depend on readings received from manned and unmanned tide gauges located throughout the Pacific, which would show whether or not the sea had been dangerously disturbed: if not, the tsunami watch would usually be cancelled, but if there were indications that a tsunami had been generated, then a tsunami *warning* would be issued.

A tsunami warning (the next level up from a tsunami watch) was the point at which the island population would be informed of the danger of incoming waves through the Emergency Broadcast System, as well as through a network of outdoor sirens installed throughout the coastal towns and villages. The sirens

Tsunami transit times across the Pacific Ocean.

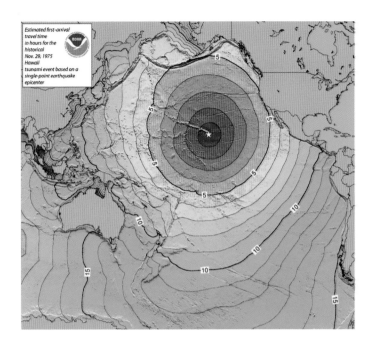

Estimated first-arrival travel time in hours for the historical Nov. 29, 1975 Hawaii tsunami event based on a single-point earthquake epicenter

would signal the immediate evacuation of low-lying areas, with hospitals placed on emergency standby, and boat-owners encouraged to move their vessels beyond the 100 fathom line.

Four years after its installation, the new warning system received its first practical test. On the morning of 4 November 1952 a powerful submarine earthquake (magnitude 8.2), shook the Kamchatka Peninsula, some 1,000 km (620 miles) west of the Aleutians. When alarms attached to seismographs in Hawaii went off just after 5 a.m., a tsunami watch was announced within the hour. Given the distance from the epicentre, any waves generated would (it was predicted) reach Hawaii by 1:30 p.m., allowing just over six hours for a coastal evacuation to take place. Meanwhile, 6-m (20-ft) waves had already begun crashing into the Kamchatka coast, and tide gauges off Alaska were registering violent movements in the water. As soon as this secondary data was in, confirming that a tsunami had indeed been generated, a tsunami warning was given out across Hawaii where, just after 1.30 p.m. – exactly as predicted – the first wave reached Hilo Bay, where it caused widespread structural damage

Tsunami warning siren, Hawaii. A familiar sight (and sound) in Pacific coastal regions, but depressingly likely to be ignored when they go off.

along the evacuated shoreline. Boats were wrecked, freight destroyed at the docks, and bridges washed away by waves of nearly 4 m (13 ft) in height. The warning system, however, had worked – just as it had in 1933 – and not a single life was lost, in spite of the fact that a party of tourists on Waikiki Beach had apparently headed towards the water instead of away from it, disoriented by the blaring siren.

Five years later, on 9 March 1957, the warning system passed its second test, when a dawn alarm call woke duty seismologists across the Pacific. At 4.22 a.m. an Aleutian Trench earthquake (magnitude 8.3) sent large waves smashing into adjacent islands, at which a general Pacific tsunami warning was issued. As in 1952, all island hospitals were placed on emergency standby, while police and the military began the work of evacuating low-lying areas. And, as before, the incoming waves caused some structural damage throughout the Hawaiian islands, but again no lives were lost, not least because of the sizeable area of downtown Hilo that had not been rebuilt after 1946, leaving a wide, grassy flood-plain to serve as a tsunami buffer zone.

But there was a dangerous chink in Hawaii's surrounding armour, for while seismographic information remained easily available, alerting scientists to earthquakes occurring anywhere in the Pacific, the vital oceanographic information was less easy to obtain. Most of it was sourced from tide gauges situated off a few rich northern Pacific coasts – Russia, Japan and the USA – while Pacific coastlines to the south of Hawaii, especially those of Chile and Peru, remained poorly equipped. As a result, a possible tsunami generated by a Central or South American earthquake could not be corroborated by tidal reports, a situation that led to scientists erring on the side of caution and issuing tsunami warnings in every case, when tide gauge data would usually have led to a tsunami watch being cancelled. A strong Mexican earthquake in 1958, for example, had prompted a tsunami warning on Hawaii, complete with a full-scale evacuation, but when no tsunami arrived, the evacuees were unimpressed by the seismologists' explanations.

Tsunami buffer zone, Hilo, Hawaii. A wide, grassy area between the ocean and the town that was set aside in the wake of the 1946 tsunami.

It is a strange quirk of human nature that we react almost as badly to false alarms as we do to unpredicted events, and it only

takes a couple of unnecessary scares to undermine the credibility of an entire warning system. In June 2007, for example, villagers near Lhok Nga, in Indonesia's Aceh province – the area worst affected by the Boxing Day tsunami – vandalized a warning siren that had gone off by accident, three days after a siren in Banda Aceh had also gone off accidentally, causing widespread panic and disruption. Though the majority of false alarms are now due to mechanical faults rather than inaccurate predictions, the effect on people's confidence is the same.[5]

It was just such a loss of confidence that led to renewed disaster in Hawaii. On 22 May 1960, an unusually powerful earthquake off the Chilean coast was detected by seismologists, who set about issuing an immediate tsunami warning. But when the sirens went off that evening, most of the coastal population of Hawaii, remembering the recent Mexican debacle, decided that they had heard it all before. Many people chose to stay at home, while others gathered along the shore to see whether a wave was really on its way, ignoring the Nietzschean dictum that 'you don't look at a tidal wave; a tidal wave looks at you'.[6] They were not entirely to blame, since a confusing change to the evacuation procedure had been introduced a few months before. Where previously the warning sirens had been sounded in three distinct stages, the last of which meant 'evacuate now', the new system featured only a single siren – the signal to begin immediate evacuation. But this new procedure was widely misinterpreted, and after hearing the night's first (and only) siren sounded at 8.30 p.m., many of those who did intend to leave began to get ready, while waiting for the final signal.

But there was no final signal, and just after 1 a.m. a devastating 10-m (32-ft) wave crashed into Hilo Bay where it claimed the lives of 61 people, badly injured 282 more and wiped out most of the streets that had been rebuilt after 1946. At magnitude 9.5, the originating earthquake remains the most powerful ever recorded, and the corresponding strength of the tsunami can be gauged from the fact that the steel parking meters in downtown Hilo were bent to the ground by the force of the retreating wave.

On a Tripoli beach, bathers mistake a marine mirage for an incoming tsunami, from an illustration in *La Domenica del Corriere*, 9 August 1936. Over time, such false alarms do much to diminish the effectiveness of warning systems.

The warning system fails: just before midnight on 22 May 1960, sightseers in Hawaii ignore the sirens and gather to wait for the long-range tsunami from Chile. 61 people lost their lives in Hilo, as did a further 142 in Japan. In Chile itself, some 6,000 people were killed by the tsunami-genic earthquake.

A row of steel parking meters bent to the ground by the force of the retreating wave. Hilo, Hawaii, 23 May 1960.

Wave action in Japan during the May 1960 tsunami: 142 people were killed in Japan by the long-distance tsunami from Chile, a day after the originating earthquake.

The warning system had failed, just as it would fail in Japan eight hours later, where 142 people were killed by the same long-distance tsunami, many of whom, it was reported, had walked out to collect stranded fish in the wake of the sea's withdrawal, or simply out of a fatal desire to see what a tsunami was like.

It was clear that the warning system needed to regain its authority, and the first problem to be fixed was the system of siren calls. Instead of the ambiguous single alert, which had caused so much confusion that evening, a new 'attention/alert' signal ('a steady three-minute siren tone', as is it described in the Hawaiian telephone book) would be sounded at least three hours before the estimated wave arrival time, and repeated at regular intervals. Meanwhile, detailed evacuation instructions would be broadcast over the radio, in which people would be encouraged to walk inland rather than drive, as a recent tsunami warning had resulted in gridlock on the coastal highway, leaving hundreds of cars stranded in the path of the incoming wave.

It was also clear that more tide gauges were needed near southern Pacific coasts, an outcome achieved after 1968 when, under the auspices of the United Nations' Intergovernmental Oceanographic Commission (IOC), all Pacific nations joined the warning system family. Today, there are two main warning centres

The site of the former Japanese-Hawaiian township of Shinmachi, rebuilt after the 1946 tsunami, but left bare after its subsequent destruction by the tsunami of 22–23 May 1960.

Coastal residents on Oahu collect stranded fish during the recession of the tsunami of 1957. But even after a lengthy interval, there is no guarantee that another wave will not be on its way.

that monitor the Pacific: the West Coast and Alaska Tsunami Warning Center in Palmer, Alaska, which covers Alaska, British Columbia, Washington, Oregon and California, and the original Pacific Tsunami Warning Center on Hawaii, which serves as the regional warning centre for the Pacific islands as well as the international warning centre for long-range and trans-oceanic tsunamis. The centres monitor seismic, tidal and sea-level information from a wealth of sources across the region, and use satellite and other technology to issue tsunami watches and warnings.

But technology on its own can offer little protection. If a major tsunami occurs less than once per generation, there is plenty of time for lessons to be forgotten. As has already been seen, disaster amnesia and warning fatigue are fatally common conditions, even among the most hazard-prone communities. The clearest lesson from the 1960 tsunami was that the real weakness of the warning system was not technological, but psychological. People were simply not scared enough of tsunamis, in spite of the islands' well-documented history of seismic assault from the sea. In fact, many Hawaiians seem transfixed by tsunamis, especially the young, who not only lack direct experience of their dangers, but are immersed in a long-established surfing culture, with its daredevil mythologies of riding the Ultimate Wave.

This new hazard became increasingly apparent during the 1980s and '90s, when any announcement of a tsunami watch or warning would send dozens of youngsters down to the shoreline with surfboards under their arms, ready to re-enact Holoua's legendary ride. I wonder how many Hawaiian surfies hoping to catch an 'Off the Richter' (slang for a particularly awesome wave) are aware of the horrible aptness of the term? A full-scale coastal

Tsunami Hazard Zone sign bearing the simple instruction to head inland in the event of an earth tremor.

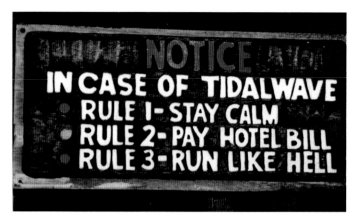

Sign in the lobby of Uncle Billy's Hilo Bay Hotel, Hawaii; testament to the bravado with which the Hawaiians continue to respond to the ever-present dangers of the sea.

evacuation on 7 May 1986, dubbed 'Waveless Wednesday' by a disappointed local press, had seen a column of traffic heading *towards* Waikiki Beach for 'tsunami party time!', in Walter Dudley's words, while later alerts generated a series of bizarre responses, including a group of mainland tourists who demanded to know what this 'Salami Warning' was all about, and the fisherman seen casting his line over the Waikiki breakwater, serenely ignoring the frantic announcements of the civil defence helicopter hovering overhead. And, at precisely the moment that the wave had been predicted to make landfall at Honolulu, 'a woman with dark flowing hair, wearing a purple muumuu, crossed the street, and calmly marched into the waves. She gazed toward the horizon while reverently touching the water and then motioned toward her heart.'[7] Luckily for all these people, the tsunamis that did arrive were knee-high or less, a fact that did little to alleviate the frustration of the civil defence authorities. As Doug Carlson, a Honolulu-based tsunami blogger, pointed out in his post for 16 November 2006, after yet another tsunami alert had sent surfers hurrying down to Waikiki Beach, 'the loonies who want to "ride a tsunami" are probably beyond hope' and 'officials may have to acknowledge that they can't change those people'.[8] Even after the recent tragedies of 2004 and 2011 it still cannot be reiterated often enough that tsunamis are completely unsurfable.

Because the culture of non-compliance during tsunami alerts is bolstered every time there is a 'false alarm' – a phrase that dismays mitigation professionals, since even a one-centimetre-high tsunami is still a tsunami, and can easily shoal into a monster given the right conditions – efforts to fine-tune the detection system are ongoing, with the aim of keeping unnecessary evacuations to a minimum.

The main improvement has been to replace the first generation of near-shore tide gauges with electronic seabed sensors – known as bottom pressure recorders (or BPRs) – that are able to monitor long-period waves passing through deep water. The problem with the old harbour-side tide gauges was that their proximity to coastlines tended to contaminate the tsunami signal, as well as the fact that they would often be destroyed by the very thing they were set up to measure. They were also slow at relaying information, but the use of satellite technology to collect real-time data transmitted from surface buoys moored directly above the seabed sensors has reduced alert times to a matter of minutes. The buoys, which are part of the National Oceanic and Atmospheric Administration (NOAA)'s Deep-ocean Assessment and Reporting of Tsunamis (DART) system, must be replaced every year, while the bottom pressure recorders are replaced every two years, making the DART system expensive to maintain. Until recently, only the Pacific Ocean was covered, with a chain of 41 buoys located above known sub-duction zones, but following the 2004 Indian Ocean tsunami, the Pacific Tsunami Warning Center has extended its remit to include the Indian Ocean, South China Sea and the Caribbean, each of which is in the process of installing or extending their own tsunami warning systems under the guidance of the Intergovernmental Oceanographic Commission (IOC).

But even after the first DART buoys had been installed in the Indian Ocean, disaster continued to strike. In July 2006 a tsunami generated off the south coast of Java killed nearly 700 people: it emerged that the two DART buoys installed off the Sumatran coast had been removed from the sea, awaiting repairs. Given that these buoys cost around $300,000 each, and require at least $125,000-worth of annual maintenance per unit, tsunami preparedness has proved a costly undertaking for nations such as Indonesia, which

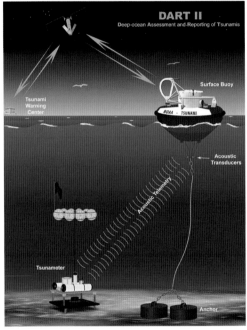

Tsunami detection system: A Deep Ocean Assessment and Reporting of Tsunamis (DART) buoy, with satellite transmitter:
1 Recorder on seabed measures pressure and sends data to buoy.
2 Buoy also detects changes in the sea level and motion.
3 Information is transmitted via satellite to ground stations which assess risk of tsunami.

has made slow progress in extending its coastal warning system. There are still only 22 detection buoys to monitor the 6,000 inhabited islands of the sprawling archipelago. The IOC, however, is continuing to expand the UN-funded warning system, which, by 2014 is set to comprise 160 seismological stations, numerous additional DART buoys and sensors, and a network of automated warning sirens ranged along the region's most populous coastlines. (Though, in light of the vandalism of a malfunctioning siren in

Aceh province, Indonesia, in June 2007, maintenance will always remain a high priority.) Regular practice drills now take place every two years: in May 2013 a Pacific-wide exercise named PacWave 13 tested all existing tsunami monitoring equipment across the region, as well as a number of new procedures including instant text messages sent from the Pacific Tsunami Warning Center, designed to deliver bespoke local warnings rather than general region-wide alerts.

But, as always, the technology can only be part of the picture. In September 2009, nearly 200 people were killed across

The DART II system consists of anchored sea-floor pressure recorders which send real-time information on wave activity to surface buoys, which then transmit the information to the Pacific Tsunami Warning Center via satellite. The technology is expensive to maintain, however, and out of the price range of many governments across the developing world.

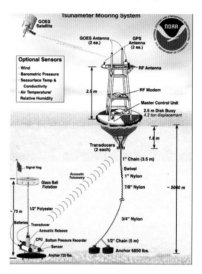

All current DART buoy locations are shown on this map.

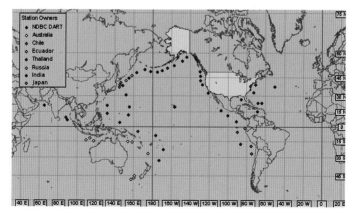

Samoa and Tonga by a powerful tsunami generated by a local undersea earthquake. Tsunami warnings had been issued that morning, some via automated text message, and evacuations did take place; but many of those killed on Samoa were caught by the second, larger wave, as they returned to the beaches to pick up fish that had been washed ashore by the first, having assumed that the second siren was a false alarm. Such episodes exemplify the widespread erosion of tsunami knowledge among island communities, and highlight the need for continued education, a process that best begins in the classroom. One of the clearest illustrations of this is the story of the ten-year-old British schoolgirl, Tilly Smith, who was on holiday with her family on Phuket, Thailand, during Christmas 2004. As soon as she saw the withdrawal of the water on the hotel beach that Boxing Day morning she knew exactly what it meant, because two weeks earlier she had been shown a video of one of the Hilo tsunamis during a school geography lesson. 'I was having visions from the Hawaiian videos that I had seen two weeks before', she recalled, and her loud insistence that a tsunami was coming persuaded the hotel staff to evacuate the beach: the only one on Phuket island where nobody was killed.[9] In a similar episode a dock worker in a remote Indian coastal area, who had recently watched a National Geographic TV documentary on tsunamis, had recognized the sea's retreat for what it was, and shouted to his co-workers to run inland. His actions saved several hundred lives.[10]

Such stories reinforce the fact that, in spite of all the expensive technology at our disposal, we should never underestimate the lifesaving potential of someone running away from the water shouting 'tsunami!', but that can only be achieved through education, whether formally in schools, or informally in the kind of cautionary folk tales that appeared in chapter Three, and that often provide the first line of defence against the ever-present dangers of amnesia.

It was the same need for greater public awareness that led to the founding of the Pacific Tsunami Museum in Hilo, Hawaii, in 1994, its primary mission being to provide tsunami education

The tsunami strikes a beach resort on Phuket island, Thailand, on Boxing Day morning, 2004. The beach – evacuated after 10-year-old Tilly Smith raised the alarm – was the only beach on the island where nobody was killed.

programmes across the Asia/Pacific region. Aptly enough, the museum's first board meeting was postponed due to a tsunami alert prompted by an 8.1 magnitude earthquake off the Kuril Islands, a timely reminder of the very need for such an institution. Two years later, the unexpected outpouring of tsunami recollections that marked the fiftieth anniversary of the 1946 disaster made it clear that the museum had another, equally important, function, as a repository for Pacific tsunami memories, and as a place where generations of survivors could meet to 'talk tsunami', with an annual Tsunami Story Festival taking place every April.[11] The museum opened in 1997, housed in a solid former bank building that survived the inundations of 1946 and 1960. From the roof a tsunami-cam keeps perpetual watch over the shoreline, ready to spot the telltale withdrawal of the water that Hawaiians call a *kai mimiki*. The islanders are determined that they will not be caught out again.

The BayCam on the roof of the Pacific Tsunami Museum, Hilo, keeps perpetual watch over Hawaii's northeast coast, waiting for the next tsunami. The live stream can be viewed online at www.tsunami.org/hilobaycam.html.

In this respect, Hilo is an exception: historically, most towns and cities have tended to suppress their memories of disaster, whereas much of downtown Hilo has been transformed into a continuous memorial, with bay-front parks studded with monuments to the city's tsunamis, including the eerie Waiakea town clock, stopped at 1.04 a.m. on 23 May 1960. Nearby, a sinuous wall of black lava, built in memory of the 1946 tsunami, fronts a large landfill plateau that was constructed above the 1960 high-water mark by army engineers. 'As a gesture of confidence', notes Mike Davis, 'state and federal offices were relocated here, along with the aptly named Kaiko'o ("violent seas") mall.'[12]

Hilo's memorial parks serve another function, however, which is to discourage redevelopment along the tsunami-prone shoreline. A similar dual intention lies behind an ongoing proposal to plant around 16,000 cherry trees at 10-m intervals along a section of Japan's Sendai coast: one for every victim of the 2011 tsunami. Like Japan's ancient tsunami stones, with their lapidary warnings, the memorial trees are intended to act as graceful disincentives to anyone planning to rebuild too close to the wave-shattered shore.

Tsunami memorials

In March 2012, exactly a year after the Japan tsunami, the rusting hulk of the *Ryou-Un Maru*, a 150-tonne Japanese shrimping boat, was spotted off the coast of British Columbia. It had spent the previous twelve months drifting unmanned across the Pacific Ocean in the vanguard of more than two millions tonnes of floating debris that had been washed out to sea by the tsunami. By 1 April, the vessel – whose name translates as 'fishing luck' – had entered United States waters, off the coast of Alaska, where it was declared a hazard to shipping and sunk by a coastguard gunship.

The ghost ship was only the first of thousands of 'tsunami objects' that have made their way across the vastness of the Pacific, including, most spectacularly, a 188-tonne concrete dock that washed up on a beach in Oregon in June 2012, after floating more than 7,000 km (4,350 miles) from the port of Misawa on Japan's northwest coast. After sitting on the beach for nearly a month, the dock was broken up and removed, with a small section – a 12,700 kg (28,000 lb) cornerstone – reserved for display at the Hatfield Marine Science Center in the nearby city of Newport, where it stands as a memorial to the 16,000 victims of the tsunami.

There is a long history of marooned objects, usually from shipping, being transformed into maritime memorials. The

Tsunami memorial, Hilo, Hawaii, a wave-shaped wall of black lava built in memory of those who died in the 1946 tsunami.

The Waiakea town clock, stopped by the wave from Chile at 1:04 a.m. on 23 May 1960.

image of the Dutch paddle-steamer, *Berouw*, stranded 3 km inland by the waves from Krakatoa, remained a lasting icon of tsunamic fury, long after the vessel itself rusted away, until all that was left was its mooring buoy. The buoy still stands on a plinth in Lampung Bay, southern Sumatra, the island's only memorial to the Krakatoa dead. In Banda Aceh, at the northern end of the island, a 2,600-tonne power-generating ship, the *Apung 1*, has been preserved where it was dropped by the wave in 2004, more than 2 km (a mile and a half) inland. It is now an official memorial to the 170,000 Acehnese who died in the Boxing Day tsunami, complete with information plaques, a landscaped garden and even an observation deck, from where visitors can gaze both over the newly rebuilt town and out towards the distant sea.

Nearby, Banda Aceh's strikingly designed Tsunami Museum, which opened in 2009, combines the myriad functions

188-tonne concrete and steel dock, washed up onto Agate Beach, Oregon, in June 2012, having drifted 7,000 km from the tsunami-damaged port of Misawa on the northeast coast of Japan.

A sizeable barge driven 50 yards inshore by a storm surge, Funchal, Madeira, December 1926.

The *Obi-iwa* rock, a 12.5 m-high boulder deposited by the Great Yaeyama tsunami of 1771, Shimoji island, Okinawa Prefecture, Japan.

of memorial, information centre and emergency shelter, featuring an internal 'escape hill' for visitors to ascend should another tsunami strike. The town appears to be emulating Hilo, with its tsunami-centred public monuments, but these costly civic memorials have proved less popular with local residents than the area's unofficial sites of remembrance, in particular the stranded fishing boat, dubbed 'Noah's Ark', which has become a focus for the ongoing grief and distress of the largely Muslim community. The 30-m (100-ft) vessel, which reportedly saved the lives of 59 people who boarded it and rode it to safety on Boxing Day 2004, ended up wedged on the roof of a house in Lampulo Port, Banda Aceh's main fishing harbour. The house's owners attributed their own survival to the miraculous descent of the boat, which remains revered as a godsend, lovingly painted and maintained by local volunteers, in contrast to the menacing

Acehnese women pray beneath a wooden fishing vessel, left stranded on the roof of a house in Lampulo village, Banda Aceh, after the Indian Ocean tsunami. The vessel, known locally as 'Noah's Ark', has become an unofficial tsunami memorial.

A child looks at a diorama of the incoming wave in the Aceh Tsunami Museum, Banda Aceh, Sumatra. The museum was built as a memorial to the victims of the Boxing Day disaster, and functions as an education centre as well as an emergency shelter in the event of future tsunamis.

hulk of the state-owned *Apung I*, from beneath which some of the bodies of those it crushed on its journey inland have yet to be recovered.

Such ambivalence over disaster memorials is understandably common. In March 2011 a 330-tonne fishing trawler, the *Kyotoku Maru No. 18*, was carried more than a kilometre inland by the wave before coming to rest amid the wreckage of the devastated port city of Kesennuma, Miyagi Prefecture. The trawler soon became a focus for offerings and prayers, an unofficial memorial to the 837 townspeople killed by the tsunami. The *Kyotoku Maru* was illuminated during the sombre anniversary services of 2012 and 2013, and had come to serve a similar function to the ancient tsunami stones with which this book began, a visible encouragement to all who saw it to 'remember the calamity of the great tsunamis'.

But, two and a half years after its stranding by the wave, the survivors of Kesennuma decided that enough was enough, finding that their iconic ship was no longer a consoling memorial, but a painful reminder of the disaster. On 8 August 2013, following a city-wide consultation, they voted for it to be taken away and scrapped, a casualty of the ever-present tension between the lifesaving need to remember and the self-preserving need to forget.

It is a tension that, in manifold ways, has shaped every episode in this book.

The 330-tonne fishing trawler *Kyotoku Maru No. 18*, stranded more than a kilometre inland in the devastated port city of Kesennuma, Miyagi Prefecture, Japan.

TSUNAMI TIMELINE

This list includes all tsunamis mentioned in this book, as well as some of the many others that have struck the world's coastlines over the centuries.

Date	Location	Cause and impact
c. 6100 BC	Norwegian Sea	Massive underwater landslides ('the Storrega slides'). Wave height unknown; tsunami deposits discovered in eastern Scotland, 80 km (50 miles) inland.
c. 1620 BC	Santorini, Greece	Eruption of Thera. Vast tsunamis causing an unknown number of deaths on Thera and Crete.
479 BC	Potidaea, Greece	Probable undersea earthquake. Unknown number of casualties.
426 BC	Malian Gulf, Greece	Earthquake. 'Huge wave' (Thucydides) claimed an unknown number of lives.
373 BC	Helike, Greece	City destroyed and sunk by earthquake and tsunami. Unknown number of casualties.
AD 365	Eastern Mediterranean	Undersea earthquake. Wave penetrated far inland. Many casualties.
563	Lake Geneva	Rockfall. Max. wave height: 8 m (25 ft). 'A large number' of casualties.
684	Hakuho, Japan	Undersea earthquake. Wave height and death toll unknown.
869	Sendai, Japan	Offshore earthquake. 'Huge waves' that penetrated *c.* 4 km inland. 1,000 casualties.
1132	Japan	Earthquake. Unknown number of casualties.
1531	Lisbon	Earthquake. Wave height unknown. Numerous casualties.
1605	Eastern Japan	Undersea earthquake. Max. wave height 30 m (98 ft). Around 5,000 casualties.

1607	Bristol Channel, Britain	Possible tsunami from undersea earthquake. Floodwaters reached 7.74 m (25 ft) above sea level. 2,000 drowned.
1611	Sanriku, Japan	Prob. undersea earthquake. Towering waves. 5,000 reported drowned.
1692	Port Royal, Jamaica	Undersea earthquake. Wave height: 2.5 m (8 ft). *c.* 2,000 casualties.
1700	Pacific-wide	Cascadia earthquake. Many drowned on the Pacific Northwest coast, as well as in Japan.
1707	Hōei, Japan	Offshore earthquake. Max. wave height: 10 m (32 ft). *c.* 5,000 casualties.
1741	Hokkaido, Japan	Volcanic eruption and landslide. Waves caused *c.* 1,500 deaths.
1755	Lisbon	Undersea earthquake. Max. wave height: 12 m (40 ft). *c.* 30,000–90,000 casualties.
1771	Yaeyama Islands, Japan	Undersea earthquake. Max. wave height: > 40 m (130 ft). *c.* 12,000 deaths.
1783	Calabria/Messina, Italy	A series of five earthquakes, two of which were tsunamigenic. Many casualties.
1792	Mt Unzen, Japan	Eruption and earthquake. Max. wave height: 20 m (66 ft). 10,000 killed by the tsunami, another 5,000 by an associated landslide.
1833	Sumatra	Powerful undersea earthquake. Wave height unknown. Casualties described as 'numerous'.
1835	Concepción, Chile	Undersea earthquake. Max. wave height: 7 m (23 ft). 100 casualties, as most inhabitants ran to higher ground.
1854	Tōkai region, Japan	Undersea earthquake. Max. wave height: 8.4 m (27 ft). *c.* 3,000 deaths from earthquake and tsunami.
1867	Virgin Islands	Undersea earthquake. Estimated wave height: 12 m (40 ft). 23 casualties.
1868	Hawaii	Undersea earthquake. Max. wave height: 18 m (60 ft). 47 killed by the tsunami, plus 31 others killed by a landslide.
1883	Krakatoa Volcanic eruption	Max. wave height: *c.* 40 m (130 ft). Nearly 40,000 casualties.
1888	Ritter Island, New Guinea	Volcanic flank collapse. Max. wave height: 15 m (50 ft). *c.* 3,000 casualties.

1896	Sanriku, Japan	Offshore earthquake. Max. wave height: 38.2 m (125 ft). 27,000 casualties.
1906	Colombia/Ecuador	Earthquake. Est. wave height: 5 m (16 ft). 500 casualties.
1907	Simeulue, Indonesia	Undersea earthquake. Wave height unknown; half the island population killed.
1908	Messina and Reggio, Italy	Undersea earthquake. Max. wave height: 12 m (40 ft). Casualties in the tens of thousands.
1923	Kamchatka Peninsula	Earthquake. Max. wave height: 8 m (26 ft). Caused one fatality in Hawaii.
1923	Japan	Great Kanto earthquake. Wave height: 10 m (33 ft). Several hundred casualties of the tsunami, plus *c.* 100,000 killed by the earthquake.
1929	Burin Peninsula, Newfoundland	Undersea earthquake, Grand Banks. Wave height: 7 m (23 ft). 28 fatalities.
1930	Papua New Guinea	Undersea earthquake. Large waves. Few casualties, as most inhabitants had time to run inland.
1933	Sanriku, Japan	Undersea earthquake. Max. wave height: 28.7 m (94 ft). *c.* 3,000 casualties.
1944	Tonankai, Japan	Undersea earthquake. Max. wave height: 10 m (32 ft). *c.* 1,200 fatalities.
1946	Hilo, Hawaii	Undersea earthquake, Aleutian Islands. Max. wave height: 17 m (56 ft). 165 casualties.
1946	Nankaidō, Japan	Offshore earthquake. Max. wave height: 6 m (20 ft). *c.* 1,500 casualties.
1952	Pacific-wide	Earthquake, Kamchatka Peninsula. Max. wave height: 6 m (20 ft). *c.* 2,000 fatalities on the Kuril Islands.
1957	Pacific-wide	Earthquake, Aleutian Islands. No reported casualties.
1958	Lituya Bay, Alaska	Rockfall. Max. wave height: 524 m (1,720 ft): the biggest ever recorded. 5 casualties.
1960	Pacific-wide	Undersea earthquake, Chile (magnitude 9.5, the most powerful ever recorded). Max wave height: 25 m (82 ft). 61 killed on Hawaii; 142 in Japan. Many deaths in Chile from earthquake and tsunami.
1963	Vajont Dam, Italy	Massive landslide caused megatsunami *c.* 250 m (820 ft) high. Nearly 2,000 casualties.

1964	Alaska	Good Friday earthquake. Max. wave height: 30 m (100 ft). 121 casualties from Alaska to California.
1976	Moro Gulf, Philippines	Offshore earthquake. *c.* 5,000–8,000 casualties, most from the ensuing tsunami.
1983	Honshu, Japan	Offshore earthquake, Sea of Japan. Max. wave height: 10 m (33 ft). 104 casualties (100 from the tsunami).
1992	Nicaragua	Undersea earthquake. Max wave height: 9.9 m (32 ft). 116 casualties.
1992	Flores, Indonesia	Undersea earthquake. *c.* 2,500 casualties, 80 per cent killed by the tsunami.
1993	Hokkaido, Japan	Undersea earthquake, Sea of Japan. Max. wave height: 31 m (102 ft). 197 killed by the tsunami.
1994	Java, Indonesia	Undersea earthquake, Java Trench. Max. wave height: 14 m (45 ft). 223 casualties.
1998	Papua New Guinea	Undersea earthquake and landslide. Max. wave height: 15 m (59 ft). *c.* 2,200 casualties.
2004	Indian Ocean	Undersea earthquake, Sumatra. Max. wave height: 30 m (98 ft). *c.* 230,000 casualties.
2006	Southwest Java	Undersea earthquake. Max. wave height: 3 m (10 ft). *c.* 660 casualties.
2007	Solomon Islands	Undersea earthquake. Max. wave height: 12 m (36 ft). 52 fatalities.
2009	Samoa and Tonga	Undersea earthquake. Max. wave height: 14 m (46 ft). 189 casualties.
2010	Chile	Offshore earthquake. Max. wave height: 3 m (10 ft). *c.* 525 casualties of both earthquake and tsunami.
2010	Western Sumatra	Undersea earthquake. Max. wave height: 3 m (10 ft). *c.* 400 dead.
2011	Northeast Japan	Undersea earthquake. Max. wave height: 38.9 m (128 ft). Nearly 16,000 casualties.
2013	Solomon Islands	Undersea earthquake. Max. wave height: 1.5 m (4.9 ft). Thirteen fatalities, nine from the tsunami.
2014	Chile	Undersea earthquake. Max. wave height: 2.1 m (7 ft). 5 fatalities.

REFERENCES

Preface: The Tsunami Stone

1 Anton Chekhov, 'Gusev', in *About Love and Other Stories*, trans. Rosamund Bartlett (Oxford, 2004), p. 53.

2 See Lucy Birmingham and David McNeill, *Strong in the Rain: Surviving Japan's Earthquake, Tsunami, and Fukushima Nuclear Disaster* (New York, 2012), p. 40.

1 Tsunamis in History and Memory

1 *Tsunamis Remembered: Oral Histories of Survivors and Observers in Hawai'i* (Honolulu, 2000), vol. 1 (n.p.).

2 Haraldur Sigurdsson et al., 'Marine Investigations of Greece's Santorini Volcanic Field', *Eos: Transactions of the American Geophysical Union*, 87 (2006), pp. 337–42.

3 *The Dialogues of Plato*, trans. Benjamin Jowett (Oxford, 1871), vol. II, p. 521.

4 *Herodotus, with an English Translation by A. D. Godley* (New York and London, 1920–24), vol. IV, pp. 131–3. See also T. C. Smid, 'Tsunamis In Greek Literature', *Greece & Rome*, second series, 17 (1970), pp. 100–104.

5 Klaus Reicherter et al., 'Holocene Tsunamigenic Sediments and Tsunami Modelling in the Thermaikos Gulf Area (Northern Greece)', *Geophysical Research Abstracts*, 12 (2010), p. 12,033.

6 See Doron Nof and Nathan Paldor, 'Are There Oceanographic Explanations for the Israelites' Crossing of the Red Sea?', *Bulletin of the American Meteorological Society*, 73 (1992), pp. 305–14; and Carl Drews and Weiqing Han, 'Dynamics of Wind Setdown at Suez and the Eastern Nile Delta', *PloS ONE,* 5(8): e12481. doi:10.1371/journal.pone.0012481.

7 Thucydides, *The History of the Peloponnesian War*, trans. Richard Crawley (London, 1874), p. 230.

8 Gretel Ehrlich, *Facing the Wave: A Journey in the Wake of the Tsunami* (New York, 2013), p. 132.

9 In common with most Japanese nouns, *tsunami* remains the same whether singular or plural. In accordance with Anglophone usage, however, and to avoid possible confusion, I have pluralized the word with a suffixed 's'.

10 Translation by Akitsune Imamura, in 'Past Tsunamis of the Sanriku Coast', *Japanese Journal of Astronomy and Geophysics*, 11 (1934), pp. 79–93.

11 Lucy Birmingham and David McNeill, *Strong in the Rain: Surviving Japan's Earthquake, Tsunami, and Fukushima Nuclear Disaster* (New York, 2012), pp. 39–40. See also Daisuke Sugawara et al., 'The 2011 Tohoku-oki Earthquake Tsunami: Similarities and Differences to the 869 Jogan Tsunami on the Sendai Plain', *Pure and Applied Geophysics*, 170 (2013), pp. 831–43.

12 *An Account by an Eye-witness of the Lisbon Earthquake of November 1, 1755* (Lisbon, 1985), pp. 14–15.

13 'An Account of the Agitation of the Sea at Antigua, Nov. 1, 1755', *Philosophical Transactions of the Royal Society*, 49 (1756), p. 669.

14 Cited in M. A. Baptista and J. M. Miranda, 'Revision of the Portuguese Catalog of Tsunamis', *Natural Hazards and Earth System Sciences*, 9 (2009), p. 27.

15 Cited in Tom Simkin and Richard S. Fiske, *Krakatau 1883: The Volcanic Eruption and its Effects* (Washington DC, 1983), pp. 307–8.

16 Cited in Rupert Furneaux, *Krakatoa* (London, 1965), p. 138.

17 Quoted in *The Times*, 12 December 1883, p. 10.

18 Cited in Simon Winchester, *Krakatoa: The Day the World Exploded 27th August 1883* (London, 2003), p. 276.

19 Furneaux, *Krakatoa*, p. 167.

20 Walter C. Dudley and Min Lee, *Tsunami!*, 2nd edn (Honolulu, 1998), p. 292.

21 In *Tsunamis Remembered* (n.p.).

22 Cited in Dudley and Lee, *Tsunami!*, p. 284.

23 Hugh Davies, 'Tsunamis and the Coastal Communities of Papua New Guinea', in *Natural Disasters and Cultural Change*, ed. Robin Torrence and John Grattan (London, 2002), p. 41.

2 The Science of Tsunamis

1 Francis P. Shepard, *The Earth Beneath the Sea*, 2nd edn (Baltimore, MD, 1967), p. 35.

2 Charles Darwin, *Journal of Researches into the Natural History and Geology of the Countries Visited During the Voyage of H.M.S. Beagle Round the World*, 2nd edn (London, 1845), p. 301.

3   Ibid., pp. 305–6.

4   For the questionnaire in full, see João Duarte Fonseca, *1755: The Lisbon Earthquake* (Lisbon, 2005), pp. 120–21.

5   John Michell, *Conjectures concerning the Cause, and Observations upon the Phænomena, of Earthquakes; particularly of that great Earthquake of the first of November 1755, which proved so fatal to the City of Lisbon, and whose Effects were felt as far as Africa, and more or less throughout almost all Europe* (London, 1760), p. 6.

6   Ibid., pp. 37–8.

7   Ibid., pp. 63–4. His calculation was only about 100 km (60 miles) out.

8   Ibid., pp. 50–51.

9   Ibid., p. 67.

10 Bunji Fujimoto, cited in *Tsunamis Remembered: Oral Histories of Survivors and Observers in Hawai'i* (Honolulu, 2000), vol. 1 (n.p.).

11 Shepard, *The Earth Beneath the Sea*, p. 31.

12 Ibid., p. 33.

13 G. A. Macdonald, F. P. Shepard and D. C. Cox, 'The Tsunami of April 1, 1946, in the Hawaiian Islands', *Pacific Science*, 1 (1947), p. 22.

14 Bruce Parker, *The Power of the Sea: Tsunamis, Storm Surges, Rogue Waves, and Our Quest to Predict Disasters* (New York, 2010), pp. 178–84.

15 Don J. Miller, 'Giant Waves in Lituya Bay Alaska', *Geological Survey Professional Paper 354-C* (Washington DC, 1960), p. 57. See also Walter C. Dudley and Min Lee, *Tsunami!* 2nd edn (Honolulu, 1998), p. 75.

16 Katrina Kremer, Guy Simpson and Stéphanie Girardclos, 'Giant Lake Geneva Tsunami in AD 563', *Nature Geoscience*, 5 (2012), pp. 756–7.

17 Cited in Julyan H. E. Cartwright and Hisami Nakamura, 'Tsunami: A History of the Term and of Scientific Understanding of the Phenomenon in Japanese and Western Culture', *Notes and Records of the Royal Society*, 62 (2008), p. 153.

18 Ibid., pp. 153–4.

19 A. W. Baird, 'Report on the Tidal Disturbances caused by the Volcanic Eruptions at Java, August 27 and 28, 1883, and the Propagations of the "Supertidal" Waves', *Proceedings of the Royal Society of London*, 36 (1884), pp. 248–53.

20 E. R. Scidmore, 'The Recent Earthquake Wave on the Coast of Japan', *National Geographic Magazine*, 7 (September 1896), pp. 285–9. Cited in Cartwright and Nakamura, 'Tsunami', p. 154.

21 Reprinted in Lafcadio Hearn, *Gleanings in Buddha-Fields: Studies of Hand and Soul in the Far East* (London and New York, 1897), p. 16.

The *OED* cites this as the first usage of the word in English, but as we have seen, Eliza Scidmore got there first.

22 'Seismology in Japan', *Nature*, 71 (1905), pp. 224–5. Cited in Cartwright and Nakamura, 'Tsunami', p. 155.

23 Sonia S. Anand and Salim Yusuf, 'Stemming the Global Tsunami of Cardiovascular Disease', *The Lancet*, 377 (2011), pp. 529–32. See also Mark Liberman, 'Antedating Tsunami', *Language Log*, 14 March 2011, via http://languagelog.ldc.upenn.edu (accessed 30 January 2014).

24 Cited in Gavin Kelly, 'Ammianus and the Great Tsunami', *The Journal of Roman Studies*, 94 (2004), p. 141.

25 Beth Shaw et al., 'Eastern Mediterranean Tectonics and Tsunami Hazard Inferred from the AD 365 Earthquake', *Nature Geoscience*, 1 (2008), pp. 268–76.

26 Cited in Edward A. Bryant and Simon Haslett, 'Was the AD 1607 Coastal Flooding Event in the Severn Estuary and Bristol Channel (UK) due to a Tsunami?', *Archaeology in the Severn Estuary*, 13 (2002), p. 164.

27 Cited in A. Bryant and Simon Haslett, 'Catastrophic Wave Erosion, Bristol Channel, United Kingdom: Impact of Tsunami?', *The Journal of Geology*, 115 (2007), p. 255.

28 Bryan F. Atwater et al., eds, *The Orphan Tsunami of 1700: Japanese Clues to a Parent Earthquake in North America* (Seattle, WA, and London, 2005), pp. 3–5.

29 Ibid., p. 4.

30 Cited in Ray Waru, *Secrets and Treasures: Our Stories Told Through the Objects at Archives New Zealand* (Auckland, 2012), pp. 215–22.

3 'The Hungry Wave': Tsunamis in Myth and Legend

1 Bryan F. Atwater et al., eds, *The Orphan Tsunami of 1700: Japanese Clues to a Parent Earthquake in North America* (Seattle and London, 2005), p. 20.

2 From an interview with Warren Nishimoto, collected in *Tsunamis Remembered: Oral Histories of Survivors and Observers in Hawai'i* (Honolulu, 2000), vol. 1 (n.p.).

3 Anton Alifandi, 'Saved by Tsunami Folklore', *BBC News Channel*, 10 March 2007, via http://news.bbc.co.uk. See also Bruce Parker, *The Power of the Sea: Tsunamis, Storm Surges, Rogue Waves, and Our Quest to Predict Disasters* (New York, 2010), pp. 164–5.

4 See Rupert Furneaux, *Krakatoa* (London, 1965), pp. 10–13.

5 Mary Kawena Pukui and Samuel H. Elbert, *Hawaiian Dictionary* (Honolulu, 1973), vol. 1, p. 109. See also Walter C. Dudley and

Scott C. Stone, *The Tsunami of 1946 and 1960 and the Devastation of Hilo Town* (Virginia Beach, 2000), p. 28.

6   From an information panel in the Pacific Tsunami Museum, Hilo.

7   Atwater et al., *The Orphan Tsunami*, p. 41.

8   John Goldingham, 'Some account of the Sculptures at Mahâbalipuram; usually called the Seven Pagodas', in *Descriptive and Historical Papers Relating to The Seven Pagodas on the Coromandel Coast*, ed. M. W. Carr (Madras, 1869), p. 33.

9   See 'Tsunami's Tragic Treasures', *Current World Archaeology*, 10 (2005), p. 7.

10 These and the other antipodean legends are from Edward Bryant, *Tsunami: The Underrated Hazard*, 2nd edn (Chichester, 2008), pp. 257–60.

11 Ibid., p. 259.

12 Cited in Ruth Ludwin et al., 'Dating the 1700 Cascadia Earthquake: Great Coastal Earthquakes in Native Stories', *Seismological Research Letters*, 76 (2005), p. 144.

13 Ibid., pp. 142–3.

14 From Beverly H. Ward, *White Moccasins* (1986), cited in Ludwin et al., 'Dating the 1700 Cascadia Earthquake', p. 142.

15 Bryant, *Tsunami: The Underrated Hazard*, p. 251.

16 Simon J. Day, 'Tsunami Hazard Awareness from Past Experience and the Differing Vulnerability of Indigenous and Coastal Populations', *American Geophysical Union*, Fall Meeting 2007, abstract OS 23B-04.

17 Hugh Davies, 'Tsunamis and the Coastal Communities of Papua New Guinea', in *Natural Disasters and Cultural Change*, ed. Robin Torrence and John Grattan (London, 2002), p. 37.

18 Lafcadio Hearn, *Gleanings in Buddha-Fields: Studies of Hand and Soul in the Far East* (London and New York, 1897), p. 19.

19 Ibid., pp. 20–25.

20 See Ito Kazuaki, 'Catfish and Earthquakes in Folklore and Fact', *Nipponia*, 33 (15 June 2005), via http://web-japan.org/nipponia.

21 From *De Natura Animalium*, cited in Parker, *The Power of the Sea*, p. 140.

22 Tew Bunnag, 'The Reluctant Mahout', in *After the Wave: Short Stories of Post Tsunami on the Thai Andaman Coast* (Bangkok, 2005), pp. 81–106. See also Michael Morpurgo, *Running Wild* (London, 2009), a novel for younger readers based on the same story.

23 Parker, *The Power of the Sea*, p. 182.

24 Eric Wikramanayake, Prithiviraj Fernando and Peter Leimgruber, 'Behavioral Response of Satellite-collared Elephants to the Tsunami in Southern Sri Lanka', *Biotropica*, 38 (2006), p. 776.

25 See Isabella Hatkoff, Craig M. Hatkoff and Paula Kahumbu, *Owen & Mzee: The True Story of a Remarkable Friendship* (New York, 2006).

26 Cited in Gary Stern, *Can God Intervene? How Religion Explains Natural Disasters* (Westport, CT, 2007), pp. 4, 172.

27 Cited ibid., pp. 50, 148–51.

28 Cited in T. D. Kendrick, *The Lisbon Earthquake* (London, 1956), p. 157.

29 Cited in Stern, *Can God Intervene?*, pp. 176–7.

30 Cited in Donald R. Prothero, *Catastrophes! Earthquakes, Tsunamis, Tornadoes, and Other Earth-shattering Disasters* (Baltimore, MD, and London, 2011), p. 55.

31 Cited in Stern, *Can God Intervene?*, p. 176.

32 Cited in Karen Fay O'Loughlin and James F. Lander, *Caribbean Tsunamis: A 500-Year History from 1498–1998* (Dordrecht, 2003), p. 56.

33 Douglas Myles, *The Great Waves* (New York, 1985), p. 117.

34 Robert 'Steamy' Chow, interviewed in *Tsunamis Remembered* (n.p.).

4 Tsunamis in Literature, Art and Film

1 *Tsunamis Remembered: Oral Histories of Survivors and Observers in Hawai'i* (Honolulu, 2000), vol. I, (n.p.).

2 *The Odyssey of Homer*, trans. Richmond Lattimore (New York, 1965), p. 191.

3 Apollonius of Rhodes, *The Voyage of Argo*, trans. E. V. Rieu (Harmondsworth, 1959), pp. 88–9.

4 *Ovid's Metamorphoses (Books I-II-III-IV-V)*, trans. Brookes More (Boston, 1933), pp. 15–16.

5 Kamono Chōmei, *The Ten Foot Square Hut and Tales of the Heike*, trans. A. L. Sadler (Sydney, 1928), p. 9.

6 Daniel Defoe, *Robinson Crusoe*, ed. John Richetti (London, 2001), p. 65.

7 Susan Neiman, *Evil in Modern Thought: An Alternative History of Philosophy* (Princeton, NJ, 2002), p. 1.

8 *Selected Works of Voltaire*, trans. Joseph McCabe (London, 1935), p. 1.

9 Voltaire, *Candide; or, Optimism*, trans. John Butt (Harmondsworth, 1947), p. 33.

10 Charles Dickens, 'Lisbon', *Household Words: A Weekly Journal*, 19 (25 December 1858), p. 89.

11 Sir Arthur Quiller-Couch, *Lady Good-for-nothing: A Man's Portrait of a Woman* (London, 1910), pp. 451–2.

12 R. M. Ballantyne, *Blown to Bits; or, The Lonely Man of Rakata* (London, 1889), pp. 362–78.

13 Ibid., p. 388.

14 H. E. Raabe, *Krakatoa: Hand of the Gods* (New York, 1930), pp. 319–20.

15 Paul Gallico, *The Poseidon Adventure* (London, 1969), pp. 22–7.

16 Boyd Morrison, *Rogue Wave* (New York, 2010), p. 134. The novel was published in the UK as *The Tsunami Countdown* (London, 2012).

17 Cited in Susan J. Napier, 'Ōe Kenzaburō and the Search for the Sublime at the End of the Twentieth Century', in *Ōe and Beyond: Fiction in Contemporary Japan*, ed. Stephen Snyder and Philip Gabriel (Honolulu, 1999), p. 22.

18 Sakyo Komatsu, *Japan Sinks*, trans. Michael Gallagher (New York, 1976), pp. 177–88.

19 Susan J. Napier, 'Panic Sites: The Japanese Imagination of Disaster from *Godzilla* to *Akira*', *Journal of Japanese Studies*, 19 (1993), p. 333.

20 Translation by Ryan Holmberg, in 'Manga 3.11: The Tsunami, the Japanese Publishing Industry, Suzuki Miso's Reportage, and the One Piece Lifeboat', *The Comics Journal*, 31 August 2011, via http://tinyurl.com/lt7y932. In common with many other *mangaka*, Miso donated all royalties to tsunami aid projects, while most of the manga publishing houses made their content available free of charge online (though, as Ryan Holmberg notes, most of the publishing houses were charging for their online content again within a month).

21 Quoted in Thomas Sotinel, 'Japan's Fantasy Films Act as a Buffer against the Reality of the Natural World', *Guardian*, 29 March 2011.

22 Cited in Walter C. Dudley and Min Lee, *Tsunami!*, 2nd edn (Honolulu, 1998), p. 224.

23 From *Tsunamis Remembered* (n.p.).

24 Doak C. Cox, 'The Inappropriate Tsunami Icon', *Science of Tsunami Hazards*, 19 (2001), p. 88; see also Julyan H. E. Cartwright and Hisami Nakamura, 'What Kind of Wave is Hokusai's *Great Wave off Kanagawa?*', *Notes and Records of the Royal Society*, 63 (2009), pp. 119–35.

25 William Miles Maskell, *The Earthquake at St. Thomas* (London, 1868), pp. 7–8; extracts from this appeared in many newspapers and magazines around the world, while an edited version appears in Karen Fay O'Loughlin and James F. Lander, *Caribbean Tsunamis: A 500-Year History from 1498–1998* (Dordrecht, 2003), pp. 87–92.

26 Cited in Ito Kazuaki, 'Catfish and Earthquakes in Folklore and Fact', *Nipponia*, 33 (15 June 2005).

27 Gregory Smits, 'Shaking up Japan: Edo Society and the 1855 Catfish Picture Prints', *Journal of Social History*, 39 (2006), p. 1045.

28 Anon., 'The Cyclorama', *The Spectator*, 21 (1848), p. 1252.

29 John Russell Taylor, 'An Eruptive Business', *The Times*, 31 July 1969, p. 7.

30 Stephen Keane, *Disaster Movies: The Cinema of Catastrophe*, 2nd edn (London, 2006), pp. 32–3.

31 Napier, 'Panic Sites', pp. 335–6.

32 Donna Coco, 'Making Waves', *Computer Graphics World*, 21:8 (1998), p. 72.

33 Susan Sontag, 'The Imagination of Disaster', in Sontag, *Against Interpretation and Other Essays* (New York and London, 1965), p. 225.

34 Cited in Lucy Birmingham and David McNeill, *Strong in the Rain: Surviving Japan's Earthquake, Tsunami, and Fukushima Nuclear Disaster* (New York, 2012), p. 44.

5 Living with Tsunamis: Warning Systems and Coastal Defence

1 Simon Day, *Daily Telegraph* letters page, 28 December 2004.

2 R. H. Finch, 'On the Prediction of Tidal Waves', *Monthly Weather Review*, 52 (1924), cited in Walter C. Dudley and Min Lee, *Tsunami!*, 2nd edn (Honolulu, 1998), p. 102.

3 Cited in Dudley and Lee, *Tsunami!*, p. 104.

4 G. A. Macdonald, F. P. Shepard and D. C. Cox, 'The Tsunami of April 1, 1946, in the Hawaiian Islands', *Pacific Science*, 1 (1947), p. 36.

5 See Bruce Parker, *The Power of the Sea: Tsunamis, Storm Surges, Rogue Waves, and Our Quest to Predict Disasters* (New York, 2010), p. 199.

6 Cited in Tom Simkin and Richard S. Fiske, *Krakatau 1883: The Volcanic Eruption and its Effects* (Washington DC, 1983), p. 101.

7 Dudley and Lee, *Tsunami!*, pp. 253, 272.

8 Doug Carlson, 'Hawaii Tsunami Was Real, but Some Surfers Showed How Unreal Their Reaction Was', http://tsunamilessons.blogspot.com (accessed 30 May 2013).

9 BBC News 24, 9 September 2005, 15.17 GMT.

10 See Donald and David Hyndman, *Natural Hazards and Disasters* (New York, 2006), p. 101.

11 See Mike Davis, 'Tsunami Memories: Disaster-Tourism on the Big Island', *Landfall*, 201 (2001), p. 54.

12 Ibid., p. 53.

SELECT BIBLIOGRAPHY

Atwater, Brian F., et al., eds, *The Orphan Tsunami of 1700: Japanese Clues to a Parent Earthquake in North America* (Seattle, WA, and London, 2005)

Birmingham, Lucy, and David McNeill, *Strong in the Rain: Surviving Japan's Earthquake, Tsunami, and Fukushima Nuclear Disaster* (New York, 2012)

Bryant, Edward, *Tsunami: The Underrated Hazard*, 2nd edn (Chichester, 2008)

Bunnag, Tew, *After the Wave: Short Stories of Post Tsunami on the Thai Andaman Coast* (Bangkok, 2005)

Cartwright, Julyan H. E., and Hisami Nakamura, 'Tsunami: A History of the Term and of Scientific Understanding of the Phenomenon in Japanese and Western Culture', *Notes and Records of the Royal Society*, 62 (2008), pp. 151–66

Clark, Timothy, *Hokusai's Great Wave* (London, 2011)

De Boer, Jelle Zeilinga, and Donald Theodore Sanders, *Earthquakes in Human History: The Far-reaching Effects of Seismic Disruptions* (Princeton, NJ, 2005)

Deraniyagala, Sonali, *Wave: A Memoir of Life after the Tsunami* (New York, 2013)

Dudley, Walter C., and Min Lee, *Tsunami!*, 2nd edn (Honolulu, 1998)

—, and Scott C. Stone, *The Tsunami of 1946 and 1960 and the Devastation of Hilo Town* (Virginia Beach, VA, 2000)

Ehrlich, Gretel, *Facing the Wave: A Journey in the Wake of the Tsunami* (New York, 2013)

Guidoboni, Emanuela, and John E. Ebel, *Earthquakes and Tsunamis in the Past: A Guide to Techniques in Historical Seismology* (Cambridge, 2009)

Hamblyn, Richard, *Terra: Tales of the Earth* (London, 2009)

Hough, Susan, *Predicting the Unpredictable: The Tumultuous Science of Earthquake Prediction* (Princeton, NJ, 2010)

Johns, Alessa, ed., *Dreadful Visitations: Confronting Natural Catastrophe in the Age of Enlightenment* (New York and London, 1999)

Kearey, Philip, and Frederick J. Vine, *Global Tectonics*, 2nd edn (Oxford, 1996)

Krauss, Erich, *Wave of Destruction: One Thai Village and its Battle with the Tsunami* (London, 2005)

Kusky, Timothy, *Tsunamis: Giant Waves from the Sea* (New York, 2008)

McGuire, Bill, *Global Catastrophes: A Very Short Introduction* (Oxford, 2005)

Murty, Tad S., U. Aswathanarayana and N. Nirupama, eds, *The Indian Ocean Tsunami* (London, 2007)

Myles, Douglas, *The Great Waves* (New York, 1985)

O'Loughlin, Karen Fay, and James F. Lander, *Caribbean Tsunamis: A 500-Year History from 1498 to 1998* (Dordrecht, 2003)

Officer, Charles, and Jake Page, *Tales of the Earth: Paroxysms and Perturbations of the Blue Planet* (New York and Oxford, 1993)

Parker, Bruce, *The Power of the Sea: Tsunamis, Storm Surges, Rogue Waves, and Our Quest to Predict Disasters* (New York, 2010)

Prothero, Donald R., *Catastrophes!: Earthquakes, Tsunamis, Tornadoes, and Other Earth-shattering Disasters* (Baltimore, MD, and London, 2011)

Robinson, Andrew, *Earthshock: Hurricanes, Volcanoes, Earthquakes, Tornadoes and Other Forces of Nature* (London, 2002)

Shepard, Francis P., *The Earth Beneath the Sea*, 2nd edn (Baltimore, MD, 1967)

Stephenson, Simon, *Let Not the Waves of the Sea* (London, 2011)

Stern, Richard Martin, *Tsunami! A Novel* (London, 1988)

Svensen, Henrik, *The End is Nigh: A History of Natural Disasters* (London, 2009)

Thompson, Jerry, *Cascadia's Fault: The Coming Earthquake and Tsunami That Could Devastate North America* (Berkeley, CA, 2011)

Tibbals, Geoff, *Tsunami: The World's Most Terrifying Natural Disaster* (London, 2005)

Torrence, Robin and John Grattan, eds, *Natural Disasters and Cultural Change* (London, 2002)

Tsunamis Remembered: Oral Histories of Survivors and Observers in Hawai'i (Honolulu, 2000)

Winchester, Simon, *Krakatoa: The Day the World Exploded 27th August 1883* (London, 2003)

ASSOCIATIONS AND WEBSITES

Deep-ocean Assessment and Reporting of Tsunamis (DART®)
www.ngdc.noaa.gov/hazard/DARTData.shtml

Japan Meteorological Agency
www.jma.go.jp/en/tsunami

NOAA Center for Tsunami Research
http://nctr.pmel.noaa.gov

The Pacific Tsunami Museum
www.tsunami.org

Pacific Tsunami Warning Center
http://ptwc.weather.gov

Tsunami Society International
www.tsunamisociety.org

West Coast and Alaska Tsunami Warning Center
http://wcatwc.arh.noaa.gov

World Health Organization
www.who.int/tsunami/en

ACKNOWLEDGEMENTS

It is a pleasure to thank all those who have helped in the process of researching and writing this book, notably Daniel Allen and Michael Leaman for commissioning it, and Reaktion's in-house editors. I am also grateful to the staff of the British Library, the Birkbeck Library and Senate House Library, University of London; the Pacific Tsunami Museum, Hilo, for their invaluable assistance; and to Warren Nishimoto of the Center for Oral History at the University of Hawaii at Manoa for permission to quote from the Center's unparalleled collection of interview transcripts.

Conversations with family, friends and colleagues have yielded numerous insights into the subject of tsunamis, for which my heartfelt thanks are due to Jon Adams, Giles Bergel, Gregory Dart, Markman Ellis, Angela Foster, David Hamblyn, Judith Hawley, Megan Hiatt, Mark Maslin, Michael Newton, Peter Straus, Colin Teevan and, especially, to Jo, Ben and Jessie Hamblyn.

PHOTO ACKNOWLEDGEMENTS

The author and the publishers wish to express their thanks to the below sources of illustrative material and/or permission to reproduce it:

Bigstock: pp. 75 (Hellogiant), 78 (Priyantha), 117 (Vitavalka); Bridgeman Images: pp. 14 (Ashmolean Museum, University of Oxford), 16, 17 (Look and Learn), 30 (private collection / Look and Learn), 103 (private collection / The Stapleton Collection); Corbis: pp. 23 (Mainichi Newspaper / AFLO), 44 (U.S. Navy / ZUMA / Corbis), 48, 50 (KYODO / Reuters), 81 (68 / Steve Allen / Ocean), 82, 83 (Pallava Bagla), 84 (Gunter Marx Photography), 158 (Buddy Mays), 160 (TARMIZY HARVA / Reuters), 161 (HOLTI SIMANJUNTAK / epa), 162 (Nozomi Sawada / Aflo / Nippon News); Getty Images: p. 159 top; Richard Hamblyn: pp. 142, 143, 148 top, 156, 157; courtesy of ITIC (International Tsunami Information Center): p. 148 bottom (A. Yamauchi, Honolulu Star Bulletin); © Zuza 'Miśko' Krzysik, April 2011: p. 123; Library of Congress, Washington, DC: p. 116; Mary Evans Picture Library: pp. 33, 145; Mimigu: p. 149; Namazu-tron: p. 92; National Oceanic and Atmospheric Administration (NOAA): pp. 56 (National Weather Service (NWS) Collection), 141, 152, 153; Pacific Tsunami Museum, Hilo: pp. 37, 39, 41, 53, 100, 148, 149, 150; Paipateroma: p. 159 bottom; Press Association Images: pp. 6 (Vincent Yu / AP), 25 (KATSUMI KASAHARA / AP), 155 (AP); Reuters Picture Library: pp. 8–9 (Ho New), 138 (Carlos Barria); Science Photo Library: pp. 58, 121 top (David A. Hardy), 121 bottom (Lynette Cook); University of British Columbia, Rare Books and Special Collections: pp. 68–9, 125; U.S. Geological Survey: p. 59; U.S. Navy: p. 22 (MC3 Dyan McCord); Werner Forman Archive: p. 85 (Provincial Museum, Victoria, British Columbia, Canada).

INDEX